CHEMISTRY RESEARCH AND APPLICATIONS

NITRIC OXIDE

EMERGING DEVELOPMENTS, THERAPEUTIC ROLE IN DISEASE STATES AND HEALTH EFFECTS

Chemistry Research and Applications

Additional books in this series can be found on Nova's website under the Series tab.

Additional e-books in this series can be found on Nova's website under the e-book tab.

CHEMISTRY RESEARCH AND APPLICATIONS

NITRIC OXIDE

EMERGING DEVELOPMENTS, THERAPEUTIC ROLE IN DISEASE STATES AND HEALTH EFFECTS

DOMINIC M. EVANS
EDITOR

nova publishers
New York

NOTICE TO THE READER

The Publisher has taken reasonable care in the preparation of this book, but makes no expressed or implied warranty of any kind and assumes no responsibility for any errors or omissions. No liability is assumed for incidental or consequential damages in connection with or arising out of information contained in this book. The Publisher shall not be liable for any special, consequential, or exemplary damages resulting, in whole or in part, from the readers' use of, or reliance upon, this material. Any parts of this book based on government reports are so indicated and copyright is claimed for those parts to the extent applicable to compilations of such works.

Independent verification should be sought for any data, advice or recommendations contained in this book. In addition, no responsibility is assumed by the publisher for any injury and/or damage to persons or property arising from any methods, products, instructions, ideas or otherwise contained in this publication.

This publication is designed to provide accurate and authoritative information with regard to the subject matter covered herein. It is sold with the clear understanding that the Publisher is not engaged in rendering legal or any other professional services. If legal or any other expert assistance is required, the services of a competent person should be sought. FROM A DECLARATION OF PARTICIPANTS JOINTLY ADOPTED BY A COMMITTEE OF THE AMERICAN BAR ASSOCIATION AND A COMMITTEE OF PUBLISHERS.

Additional color graphics may be available in the e-book version of this book.

Library of Congress Cataloging-in-Publication Data

Nitric oxide : emerging developments, therapeutic role in disease states and health effects / Dominic M. Evans, editor.
pages cm. -- (Chemistry research and applications)
Includes index.
ISBN 978-1-63482-961-8 (hardcover)
1. Nitric oxide--Physiological effect. 2. Neurochemistry. I. Evans, Dominic M., editor.
QK898.N58N57 2015
572'.542--dc23

2015017722

Published by Nova Science Publishers, Inc. † New York

CONTENTS

PREFACE

Nitric oxide (NO) is a pleiotropic free radical messenger molecule. There is enzymatic, non-enzymatic and bacterial production of NO in the human gut. This book examines the role of NO in human gastrointestinal and hepatic physiology and disease processes. Furthermore, it has long been recognized that nitric oxide is heavily involved in cell signaling. The authors' examine the complex relationship between S-nitrosylation and cancer biology. Other chapters in this book explore the complexity of NO metabolic pathway in order to better understand its role in the interaction between parasites and host cells; the most promising nanomaterials for the development of electro-chemically or optical sensors for in vivo NO sensing and the NO pathway in Trypanosoma cruzi epimastigotes, associated with several diseases considered to be public health concerns.

Chapter 1 – Nitric oxide (NO) is a pleiotropic free radical messenger molecule. And NO previously considered to be a potentially toxic chemical. In recent years, NO has been established as a diffusible universal messenger that mediates cell-cell communication throughout the body. NO is an intracellular and intercellular messenger with important functions in a number of physiologic and pathobiologic processes within gastroenterology and hepatology, including gastrointestinal tract motility, mucosal function, inflammatory responses, gastrointestinal malignancy, blood flow regulation and hepatocellular function. Since the broad review of this topic in gastroenterology more than several years ago, a number of advances have been made in the area of NO biology and its relevance to the gastrointestinal and hepatology system. The aim of this chapter is to focus on our expanded understanding of the role NO plays in human gastrointestinal and hepatic physiology and disease processes by drawing on data from relevant in vitro and vivo as well as observational human studies.

Chapter 2 – This chapter examines the complex relationship between S-nitrosylation and cancer biology. As a double-edged sword, S-nitrosylation has been shown to be involved in malignant transformation, cell proliferation, induction of angiogenesis, and inhibition of DNA repair and apoptotic pathways. However, on the reverse edge, this post-translational modification has the ability to induce pro-apoptotic pathways and upregulate key DNA repair genes. Herein, we discuss the conflicting signals of S-nitrosylation and its potential role in the clinical management of cancer.

Chapter 3 – Nitric Oxide (NO) is a highly reactive and unstable free radical that is produced by L-arginine oxidation to citrulline mediated by the enzyme Nitric Oxide Synthase (NOS). NO has been considered a key molecule on innate immune responses against different parasites, including *Trypanosoma cruzi* (*T. cruzi*), however a careful look at the literature about the relationship of NO and *T. cruzi* immune response shows many controversial data about it. *T. cruzi* is an obligate intracellular parasite with a complex life cycle involving invertebrates and vertebrates and is the causal agent of Chagas disease, considered one of the most important public health issues in Latin America, where several million people are infected. Different escape mechanisms have been developed by parasites thanks to their long co-evolution with host cells which exhibit many regulatory mechanisms manipulated by the parasites. Regarding NO, this is not different. Therefore, it is important to understand these mechanisms in order to develop control strategies to different infectious pathologies. Two particular aspects have called our attention: first, many parasites have NOS and exhibit the ability to produce NO including trypanosomatids; and second, even if they do not manipulate the metabolic pathway of NO production in the host cell, they are able to resist to high levels of this molecule. Despite being a potent mediator molecule with multiple functional activities as extensively demonstrated in mammalian cells, the occurrence of NO in protozoan parasites and its physiological role has received much less attention. Also, the ability to resist to high levels of NO during the infectious process has not been much exploited. The use of chemical NO donors, such as sodium nitroprusside can be helpful to evaluate both the resistance of microorganisms to this molecule and the amount of NO needed to cause toxicity. This chapter aims to discuss the complexity of the NO metabolic pathway in order to better understand its role in the interaction between parasites and host cells.

Chapter 4 – The *in vivo* NO sensing is of outstanding relevance for the comprehension of the diverse physiological and pathological process in which this species is involved. Nevertheless, *in vivo* NO sensing is a complex challenge due to its own physicochemical properties, antioxidant mechanisms and possible interferences. Some electrochemical and optical sensors have already been developed for *in vivo* NO sensing. Most of the electrochemical sensors are amperometric and most of the optical sensors are based on fluorescence. New materials are being developed for new NO electrochemical or optical sensors. Due to their intrinsic properties, the nanomaterials are one of the most promising emergent materials. This chapter review the most promising nanomaterials for the development of electrochemically or optical sensors for *in vivo* NO sensing.

Chapter 5 – Some trypanosomatids are associated with several diseases which are considered important public health problems, such as leishmaniasis, presenting in several clinical forms caused by different species of the genus *Leishmania*, and Chagas Disease, caused by *Trypanosoma cruzi*. During the infection, host cells produce toxic products, but trypanosomatids have a very well developed defense system that allows them to survive and proliferate within those cells. This system is based mainly in the L-arginine metabolism, since trypanosomatids are able to produce nitric oxide (NO) through the nitric oxide synthase (NOS) enzyme that uses L-arginine as a substrate. L-arginine is also a substrate for arginase (ARG) producing urea and polyamines, leading to the synthesis of trypanothione, a well-known molecule which is part of a redox system in these parasites. The functions and interaction of these enzymes of the L-arginine pathway in the parasite could be related to their infectivity and virulence or their susceptibility to different drugs and free radicals. Furthermore, this pathway in the macrophages could determine the parasite death or its proliferation in a dependence of the intracellular availability of L-arginine.

The NO pathway has been described in *Trypanosoma cruzi* epimastigotes and is characterized by the enzyme NOS. Our research group demonstrated the NO production and characterized a cNOS from *L. amazonensis, L. braziliensis* and *L. infantum* (syn. *L. chagasi*) promastigotes and axenic amastigotes, NO production being more prominent in the latter.. The arginase gene was identified in *L. major* and *L. amazonensis. Leishmania* parasites have one arginase gene and data from studies with a mutant *L. mexicana* without arginase expression showed that these parasites were unable to synthesize polyamines besides presenting a decrease in their infectivity. However, until now no data concerning ARG have been described in *T. cruzi.* The polyamine biosynthesis in this case is obtained through another metabolic pathway, which is unique to *T. cruzi*. Recent data suggest a role of parasite L-arginine-/agamatine-derived polyamines in this process. It is important to point out the importance of polyamines and their relation with trypanothione biosynthesis, which is essential for parasite survival. As these trypanosamiases are highly pathogenic to humans and other mammals, and a neglected area of scientific research, despite being endemic to some South American countries, it is fundamental to understand their biochemistry in order to enable more specific interventions.

In: Nitric Oxide
Editor: Dominic M. Evans

ISBN: 978-1-63482-961-8

Chapter 1

NITRIC OXIDE IN GASTROINTESTINAL AND HEPATIC DISEASE AND HEALTH

Cong Dai, Yi-ning Li, Wei-xin Liu, Min Jiang and Ming-Jun Sun
Department of Gastroenterology, First Affiliated Hospital of China Medical University, Shenyang City, Liaoning Province, China

ABSTRACT

Nitric oxide (NO) is a pleiotropic free radical messenger molecule previously considered to be a potentially toxic chemical. In recent years, NO has been established as a diffusible universal messenger that mediates cell-cell communication throughout the body. NO is an intracellular and intercellular messenger with important functions in a number of physiologic and pathobiologic processes within gastro-enterology and hepatology, including gastrointestinal tract motility, mucosal function, inflammatory responses, gastrointestinal malignancy, blood flow regulation and hepatocellular function. Since the broad review of this topic in gastroenterology more than several years ago, a number of advances have been made in the area of NO biology and its relevance to the gastrointestinal and hepatology system. The aim of this chapter is to focus on our expanded understanding of the role NO plays in human gastrointestinal and hepatic physiology and disease processes by drawing on data from relevant in vitro and vivo as well as observational human studies.

1. THE SOURCES OF NO IN THE GASTROINTESTINAL TRACT

There is enzymatic, non-enzymatic and bacterial production of Nitric oxide (NO) in the human gut. Two constitutively expressed calcium dependent isoforms of nitric oxide synthase (NOS) are responsible for the enzymatic production of small amounts of NO: neuronal NOS (nNOS, type I), and endothelial NOS (eNOS, type III) [1]. These isoforms are primarily regulated by intracellular calcium, via $Ca2^{+}$ activated calmodulin. Inducible isoform (iNOS, type II) is not regulated by calcium, because activated calmodulin is inserted at the time of synthesis [2-4]. This isoform produces large amounts of NO for a limited period of time. The induction of iNOS usually occurs in states of inflammation and immune activation.

In the gut, inflammatory, epithelial, endothelial and neuronal cells can express iNOS. There are several NOS-independent mechanisms of NO formation. For example, xanthine oxidoreductase is an enzyme that under hypoxic conditions can produce NO by reduction of nitrate (NO_3^-) and nitrite (NO_2^-) [5, 6]. NO can also be formed from dietary nitrate which in the oral cavity is reduced by bacterial reductases to nitrite. This nitrite can be acidified in the gastric lumen yielding NO gas. NO production from the reaction of hydrogen peroxide with arginine is another example of non-enzymatic NO production. Finally anaerobic bacteria in the colon produce NO, using nitrite and nitrate as substrates.

2. NO IN GASTROINTESTINAL MOTILITY

Motility of the gastrointestinal tract is directly controlled by enteric inhibitory and excitatory motor neurons that innervate the smooth muscle layers. Distension of the gastrointestinal tract occurs by a food bolus stimulus [2, 7-10]. This distension is detected by local enteric afferent neurons. These neurons project to interneurons, which send their stimulus either upstream (orally) or downstream (aborally). Upstream interneurons are stimulatory and induce contractions via the neurotransmittors acetylcholine and substance P. Downstream interneurons are inhibitory and act via the neurotransmittors somatostatin, γ-aminobutyric acid (GABA). These inhibitory interneurons regulate other interneurons that use endogenous opiates, vasoactive intestinal peptide (VIP) and NO to inhibit muscular contraction. About 50 % of the nerves in the enteric nervous system contains nNOS. These nerves are located in the myenteric plexus and muscle fibers.

Bult and colleagues were the first to demonstrate that NO is the most important non-adrenergic, non-cholinergic (NANC) inhibitory neurotransmitter in the gastrointestinal tract [11]. Inhibitory motor neurons mediate receptive and accommodative relaxations and control the opening of sphincters. Impaired NO release is observed in diseases with non-relaxing sphincters or bowel segments like achalasia, infantile hypertrophic pyloric stenosis and Hirschprung's disease. And nNOS gene therapy may perhaps in the future become a new treatment options. Topical NO donors have already been used in patients undergoing endoscopic retrograde cholangiopancreaticography (ERCP) to relax the sphincter of Oddi and inhibit duodenal motility. Isosorbide dinitrate (ISDN) ointment is locally applied to relax the anal sphincter in order to heal anal fissures. Furthermore, ISDN tablets are used to treat oesophageal spasms. Transient LES relaxation (TLESR) are important in gastroesophageal reflux disease (GERD). Intravenous infusion of the NOS inhibitor N-monomethyl-Arginine (L-NMMA) in healthy volunteers caused a decrease in the gastric distension-triggered TLESR's and an increase in oesophageal peristaltic amplitude and velocity. This indicates that inhibition of NO might be of benefit for patients with GERD. The contractile activity of the gastrointestinal tract has an intriguing pattern, which is known as the migrating motor complex (MMC). This MMC has three phases of activity with minimal contractile activity in the fast state (phase I), mixing contractility in the fed state (phase II) and high amplitude peristaltic contractions (phase III). Several studies have shown that inhibitors of NOS initiate premature phase III contractions whereas NO donors disrupt the MMC. Therefore, selective inhibition of nitric oxide synthase could be a treatment option in patients with bacterial overgrowth due to an impaired phase III activity. Also a toxic megacolon in patients with ulcerative colitis is propably to a certain extent caused by overproduction of NO by highly induced iNOS in the colonic smooth muscles. Selective iNOS inhibition could be a treatment strategy in this life threatening conditio. Several motility disorders like chronic intestinal pseudo-obstruction and even constipation might relate to the enteric NO system, thereby suggesting new pharmacological treatment options.

3. NO in Gastrointestinal Secretion and Absorption

In the gut lumen, NO has a half life of less than 6 seconds, and is rapidly converted into nitrite and nitrate in the presence of oxygen and water. It is highly diffusible in water, lipids, and air and it freely traverses cell membranes and passes into adjacent target cells. NO is involved in the intestinal water transport by acting directly on the epithelium and bloodflow or indirectly by stimulating neuronal reflexes, and releases of, or interactions, with other agents [3]. For example, NO activates soluble guanylate cyclase and this results in cGMP generation, a potent activator of intestinal secretion. NO can also induce vasoactive intestinal polypeptide (VIP) an important neurotransmittor in secretomotor neurons. Furthermore, NO causes an increase of prostaglandin (PG) E2 production, a known secretory molecule. Apart from indirect effects on secretory molecules, NO may also exert direct secretory effects by opening of chloride channel. NO is one of the mediators of the intestinal secretion and laxative induced diarrhoea induced by castor oil, magnesium sulfate, and anthraquinone containing laxatives such as senna and cascara, as well as the diphenylmethanes: phenolphthalein and bisacodyl [12]. Bile acid infusion in the left colon induces NO generation suggesting that NO is also involved in bile salt induced diarrhoea. Patients with collagenous or lymfocytic colitis produce watery diarrhoea in the absence of epithelial cell damage. High levels of NO gas in the gut lumen of these patients suggest a role of NO in inflammation induced diarrhoea. Topical administration of the NOS inhibitor *Ng*-monomethyl-L-arginine (L-NMMA) reduced fluid secretion in patients with collagenous colitis. In contrast NO can also reduce fluid secretion as demonstrated in Clostridium difficile and cholera toxin induced diarrhoea. Early studies with NOS inhibitors already showed that NO promotes absorption under basal conditions. Interestingly the secretory effect of the NOS inhibitor *NG*-nitro-L-arginine (L-NAME) could be reversed by loperamide a known antidiarrhoeal opioid. The mechanisms of the proabsorptive actions of NO are not fully understood but may involve the opening of basolaterally located potassium channels in enterocytes. In summary it seems that NO can act both as a secretagogue and an absorbagogue depending on the concentrations, local circumstances and on the site of delivery.

4. NO in Gastrointestinal Injury, Repair, Carcinogenesis and Apoptosis

The gastrointestinal tract is remarkably protected against damage by ingested irritants (food components, alcohol, drugs); endogenous secretions (acid, bile, proteolytic enzymes) and potentially harmful actions of enteric microbes [5, 13, 14]. The mucosal barrier is composed of a rapidly replaced monolayer of intestinal epithelial cells (IEC) and non-specific agents such as lysozyme, acid and mucus. NO is important in maintaining mucosal integrity by several mechanisms. A continuous supply of blood to the gastrointestinal mucosa is vital during periods of injury. Especially eNOS derived NO plays an important role in the calcitonin gene-related peptide (CGRP) mediated, reactive hyperaemic response to mucosal injury. Studies with irritants like ethanol, hydrochloric acid or the vasoconstrictive agent endothelin showed reduced gastric mucosal damage when NO donors were administered while gastric mucosal damage was increased when NO scavengers or NOS inhibitors were given. These findings spurred the development of adding NO donating groups (nitroxybutyl) to known ulcerogenic non steroidal anti-inflammatory drugs (NSAIDs). Indeed, NO-NSAIDs did not cause the gastric mucosal injury normally associated with NSAIDs and could even protect against gastric mucosal injury associated with endotoxin or hemorrhagic shock. Apart from increasing mucosal blood flow, the protective actions of NO may also involve reduction of leucocyte adherence, inhibition of mast cell activation, inhibition of Th1 type cytokines by inactivation of cytokine processing, and stimulation of gastric mucus secretion. In addition to its protective actions in response to injury, NO is also a modulator of mucosal repair. Konturek et al., demonstrated retarded gastric ulcer healing by NOS inhibitors and accelerated healing by the NO donor glyceryl trinitrate. Delayed ulcer healing was also found in iNOS knockout mice with acetic-acid induced colitis. Studies with cutaneous wounds demonstrated that NO enhances collagen production by fibroblasts. Transfection with the gene for iNOS into the wound of the skin had beneficial effects. Angiogenesis is important in both wound repair and carcinogenesis. NO derived from eNOS expressed in mammary tumor cells promoted tumor growth and metastasis by stimulation of tumor cell migration, invasiveness and angiogenesis. However, in another study with colon carcinoma cells the presence of eNOS inversely correlated with their metastatic potential. In this study the absence of NO production in metastatic cells induced platelets aggregation that increased the adhesion of metastatic cells to the vascular endothelium and promoted their successful implantation. Precancerous lesions such as gastric intestinal metaplasia and colonic polyps express iNOS. Selective inhibition of iNOS showed a reduced development of aberrant crypt foci indicating that specific iNOS inhibition could be a chemopreventive strategy for colon cancer. Early studies concerning the role of NO in carcinogenesis were focused on its potential to nitrosate (addition of NO^+) amines, including those in DNA, forming nitrosamines. Nitrosamines can lead to direct mutations or to the generation of carcinogens. NO may also potentiate DNA damage by inhibition of DNA repair mechanisms. The capacity of NO to induce apoptosis was first appreciated by Albina, who showed that NO caused apoptosis in macrophages. Accumulation of the tumor suppressor protein p53 has been described as an essential and early indicator of NO-induced apoptosis. Although high concentration of NO is a well-established cytotoxic and pro-apoptotic effector molecule, there is growing evidence that low concentrations of NO can also inhibit apoptosis. First, NO can oxidize intracellular reduced glutathione and thereby change the antioxidant levels within the cell, resulting in oxidative or nitrosative stress. This stimulates the induction of heat shock proteins HSP32 and 70, which protects cells from apopoptic cell death. Second, caspase proteases contain a single cysteine at the catalytic site that is essential for activity. Suppression of the effector caspase-3 activity by S-nitrosylation of the essential cysteine cys163 can block apoptosis 66. Finally, NO may inhibit cytochrome C release from mitochondria by inhibiting bcl-2 cleavage. Thus, NO may act as a bifunctional regulator of apoptosis with inhibition of apoptosis in case of oxidative stress as present in mucosal injury and induction of apoptosis in carcinogenesis.

5. NO in Gastrointestinal Infection and Inflammation

Despite intestinal barriers there is a constant interaction between antigens in the gut lumen and the mucosal immune system. The presence of inflammatory cells in the gut mucosa reflects this "tolerant" mucosal immune response against the normal gut content. NO plays an important role in both barrier and immune function [15-22]. Inhibition of NO with *NG*-nitro-L-arginine (L-NAME) resulted in a rapid increase of trans-epithelial movement of the small molecule ^{51}Cr-EDTA (MW 330) indicating increased permeability. The increased permeability for small molecules occurred without overt injury and could be restored by exogenous NO suggesting that NO maintains an intact mucosal barrier. However, intestinal epithelial cells (IEC) exposed to high amounts of NO showed increased permeability. Induction of iNOS can yield micromolar amounts of NO for sustained periods. The induction of iNOS is mediated by the nuclear transcription factor κB (NF-κB). Various stimuli like viruses, microbial products, pro-inflammatory cytokines (TNF-α, IL-1, IL-6), T-and B-cell mitogens, physical and chemical stress can activate NF-κB. Because a combination of cytokines (IL-1, IFN-γ, TNF-α) and endotoxin (LPS) are needed for the in vitro induction of iNOS in native colon cells and in intestinal tumour cell lines these cells are believed to be relatively resistant to cytokine induced NF-κB activation. A decreased IκB kinase (IKK) activity and consequent resistance to IκB degradation is postulated as a protective response of IEC's to remain quiescent in the hostile colon environment. Indeed, many immunohistochemical studies even combined with Western blotting could not demonstrate epithelial iNOS expression in normal gut mucosa. However, Roberts et al., found iNOS mRNA and Perner et al. demonstrated iNOS protein expression in the normal colonic mucosa suggesting a low grade physiological induction of iNOS in the normal colon. Despite relative resistance of IEC's to NF-κB activation several pathogenic organisms such as Salmonella, Shigella, Listeria and Helicobacter species can activate NF-κB and induce iNOS in IEC's. NO in itself may not be toxic to bacteria, in fact certain enteric bacteria contain nitrate reductase and produce NO by their own. However, NO becomes cytotoxic when generated together with superoxide forming peroxynitrite ($OONO^-$) a powerful antimicrobial agent. Epithelial iNOS induction has clearly been demonstrated in inflammatory conditions such as; diverticulitis, inflammatory bowel disease (IBD) and celiac disease. Epithelial iNOS induction and NO production may cause increased intestinal permeability as observed in septic patients. Indeed selective inhibition of iNOS in endotoxemic rats ameliorated mucosal permeability for dextran and reduced bacterial translocation. The absence of bacterial translocation in endotoxemic iNOS knockout mice further supports a pathogenic role of epithelial derived NO in sepsis. The role of NO in IBD is less obvious. NO can react with superoxide (O2) anions yielding the toxic reactive nitrogen intermediate peroxynitrite (OONO-). In addition to oxidation reactions, peroxynitrite can nitrate tyrosine to produce 3-nitrotyrosine. Evidence for peroxynitrite mediated damage of epithelial cells in IBD was found in one study. Studies using inhibitors of NOS in experimental colitis showed little improvement; no effects, or even worse effects on colitis probably due to the lack of iNOS specificity of the inhibitors used. The first report of experimental colitis in iNOS knockout mice showed delayed healing and persistent inflammation in acetic-acid induced colitis. Studies with trinitrobenzene sulphonic acid (TNBS) induced colitis in iNOS KO mice showed resistance to TNBS colitis in one and increased inflammation in two other studies. In the dextran sulphate sodium (DSS) colitis model in iNOS KO mice showed resistance to DSS in three studies. Finally, a chronic colitis which develops spontaneously in interleukin 10 (IL-10) deficient mice, developed at the same rate and intensity in mice deficient in both the IL-10 and iNOS genes. Considering the absence of macroscopic ulcerations in the presence of large amounts of NO in patients suffering from microscopic colitis a protective role of NO is suggested. Indeed a NO donating mesalazine derivative had an additional beneficial effect on TNBS induced colitis. The reduced gastro-intestinal toxicity of NO donating non steroidal anti-inflammatory drugs (NSAID) and aspirin are in agreement with a protective effect of NO on intestinal epithelial cells. Apart from the above-mentioned beneficial effects of NO in mucosal injury, NO can also inhibit NF-κB activation. Therefore high amounts of NO may serve in a negative feedback loop to block prolonged activation of NF-κB thereby limiting chronic inflammation.

6. NO in Gastrointestinal Mucosal Protection

Gastric Mucosal blood flow and gastric mucosal Iesions. Acute gastric mucosal erosions or ulcers that may be associated with serious gastrointestinal bleeding commonly occur in the settings of severe acute medical illness or ongoing sepsis, severe trauma or burns, CNS disease, and ingestion of gastrotoxic drugs. The pathogenesis of acute gastric mucosal lesions is multifactorial, but alterations in gastric mucosal blood flow appear important in their development. Mucosal blood flow is dependent on the balance between the endothelial release of substances that enhance blood flow through vasodilator and/or antiaggregatory properties such as prostacyclin and vasodilator neuropeptides and those substances that reduce blood flow and platelet aggregation such as thromboxane and endothelin-1. Evidence indicates that NO may be an endogenous vasodilator regulating gastric mucosal blood flow and maintaining mucosal integrity and defense.

In the rat stomach, inhibition of NO synthesis reduces gastric mucosal blood flow as measured by a hydrogen gas clearing technique, indicating that endogenous NO modulates basal tone in the gastric vasculature. Increases in gastric mucosal blood flow induced by acetylcholine or bradykinin in the rat is endothelium-dependent, and NO synthase inhibitors attenuate pentagastrin-induced increases in gastric mucosal blood flow, indicating that endogenous NO plays a role in agonist-stimulated increases in mucosal blood flow as well.

Endogenous NO may contribute to mechanisms that protect against ulcerations of the gastric mucosa. Topical mucosal application of a NO solution or of glyceryl trinitrate or nitroprusside, which release NO, reduces the severity of ethanol-induced hemorrhagic mucosal damage. Intravenous administration of nitroprusside also inhibits mucosal damage. Administration of inhibitors of NO synthesis alone did not lead to acute gastric mucosal damage in the rat. However, inhibition of NO synthesis after depletion of vasodilator neuropeptides with chronic capsaicin pretreatment or after inhibition of prostacyclin synthesis with indomethacin did induce substantial mucosal injury. Another study suggests that the gastroprotective effect of sensory neuropeptides released by acute capsaicin exposure is mediated by NO. These results suggest a role for endogenous NO, interacting with prostacyclin and vasodilator neuropeptides in the regulation of gastric mucosal integrity. The mechanism underlying the mucosal protective actions of NO is not known. It seems likely that vasodilation or inhibition of platelet aggregation in the gastric microvasculature are involved, but it is possible that NO has other actions in the gastric mucosa that enhance or preserve epithelial cell function and continuity. The source of the gastroprotective endogenous NO is also not known. The vascular endothelium seems a likely source, but given the recent demonstrations of NO formation in many cell types, other sources including white blood cells, epithelial cells, or neurons should also be considered. Speculation on clinical implications. Damage to endothelial cells may be an important primary event that leads to acute gastric mucosal ulceration. The development of mucosal ulceration after endothelial damage could be related to decreased endothelial production of vasoactive mediators with protective actions such as NO. NO production does decrease after endothelial damage in coronary blood vessels. Pharmacological interventions to increase gastric mucosal levels of NO by augmenting endogenous NO production or providing an exogenous source of NO-like nitroprusside could prove effective in the prevention of acute gastric mucosal injury in high-risk settings. The gastroprotective effects of the drug carbenoxolone may be mediated in part by NO. However, the data regarding a mucosal protective role of NO are entirely from animal studies, and the role of NO in the human gastric mucosa has not been defined. Intestinal Mucosa Endotoxin-induced intestinal damage. The cardiovascular collapse and multiple metabolic derangements associated with septic shock are largely due to bacterial endotoxin, but the toxic effects of endotoxin are mediated by the production and release of biologic mediators by the host. In the rat, intravenous injection of endotoxin induces acute intestinal vascular damage, characterized by vasocongestion, plasma leakage, and hemorrhage into the intestinal lumen. Platelet activating factor and thromboxane A, are endogenous vasoactive mediators released in response to endotoxin in the rat. They appear to be important in the production of the gastrointestinal hemorrhagic lesions induced by endotoxin. Other vasoactive mediators released by endotoxin attenuate the harmful effects of endotoxin, and may be part of the host defense to endotoxic shock. Recent evidence suggests that endogenous formation of NO maintains microvascular integrity of the intestinal mucosa following acute endotoxin challenge, and endogenous NO may be one of the mediators that help protect against the harmful effects of endotoxic shock. NO and intestinal mucosal protection. In the rat, pretreatment with an inhibitor of NO synthesis enhances endotoxin-induced small intestinal damage and plasma leakage. The nitrovasodilator S-nitroso- N-acetyl-penicillamine, which generates NO, attenuates endotoxin-induced intestinal damage in this model. These results suggest that endogenous NO protects the intestinal mucosa and microvasculature against endotoxin-induced damage. The mechanisms are largely unknown. Endotoxin can stimulate the formation of oxygen radicals such as superoxide anion, which may mediate intestinal injury. NO can interact with superoxide anion to produce a less toxic species and may act as an antioxidant, which could explain how NO reduces the microvascular damage produced by endotoxin. Neutrophil adhesion to vascular endothelium and emigration from blood vessels can also be inhibited by N0, suggesting that endogenous NO may protect the intestine from endotoxin-induced damage by limiting the damaging effects of neutrophils. Speculation on clinical implications. The animal studies described above suggest that NO plays an important role in protecting the intestine from blood-borne toxins and tissue-destructive mediators. The role of NO in the protection of the human intestine from toxic mediators is unknown, but investigation into the role of NO in inflammatory conditions in which there is tissue damage or increased vascular permeability involving vasoactive mediators, toxic oxygen radicals, or neutrophils may provide important insights into the pathogenesis of idiopathic inflammatory bowel disease, infectious colitis, ischemic gastrointestinal injury, and gastrointestinal malignancy.

7. NO in Inflammatory Bowel Disease

1. Inducible Nitric Oxide Synthase Activity Studies in IBD

NO has been extensively studied as a mediator of intestinal inflammation in ulcerative colitis and Crohn's disease, often with conflicting results. Citrulline levels can be used to assess iNOS activity since citrulline is produced from arginine in equimolar concentrations when NO is liberated by NOS. Citrulline levels were reported to be increased in rectal biopsy specimens from patients with active ulcerative colitis. Incubating biopsy specimens with NG-monomethyl-L-arginine (LNMMA), an inhibitor of NOS, prevented the increase in citrulline levels. A study from the United Kingdom demonstrated that iNOS activity was 8-fold higher in the mucosa of patients with ulcerative colitis compared with controls but was not increased in samples from patients with Crohn's disease when compared with controls. Conflicting results were obtained in a study from Israel, which demonstrated 10-fold and 3.8- fold increases in iNOS activity in patients with ulcerative colitis and Crohn's disease, respectively. Nitric oxide production in patients with IBD can be assessed in other ways, including measuring luminal NO directly, measuring mucosal NO using electrodes, and measuring metabolites of NO. Colonic gas was aspirated from patients with ulcerative colitis and normal controls during colonoscopy. Using a chemiluminescence technique, the investigators found that NO levels were increased by 100-fold in the ulcerative colitis patients. Using a catheter inserted into the rectum, aspirated air analyzed using the same technique demonstrated markedly elevated rectal NO levels in children with active ulcerative colitis, ileocecal Crohn's disease, and colorectal Crohn's disease compared with controls. These levels were higher in patients with active disease compared with inactive disease. Utilizing steady state gas perfusion of the colon, a recent study found that colonic NO production in patients with ulcerative colitis is increased relative to controls, but that patients with collagenous colitis had greater increases in NO production. Because there is typically less mucosal inflammation in collagenous colitis than in ulcerative colitis, the investigators suggested that NO may not be an important mediator of inflammation. Mucosal NO production can be measured directly with NO electrodes. Japanese investigators modified these electrodes such that they could be introduced through the biopsy channel of a colonoscope. Mucosal NO was increased in endoscopically normal and abnormal colon in ulcerative colitis compared with controls. Treatment of patients with glucocorticoids decreased mucosal NO production by 53%. Nitrite, a stable metabolite of NO, can be readily measured in the urine. A 12-hour urinary collection revealed detectable urinary nitrite levels in 28 of 32 patients with active ulcerative colitis compared with 2 of 14 with inactive disease. None of the healthy controls had detectable urinary nitrite. In summary, iNOS expression and activity, and hence NO production, are generally increased in IBD. This relation appears to be most strong in ulcerative colitis compared with Crohn's disease. However, whether NO is a nonspecific marker of intestinal inflammation or a direct mediator of the inflammatory response in humans remains uncertain.

2. Immunohistochemical Studies in IBD

What cells are responsible for initiating and maintaining iNOS activity? iNOS protein expression was shown to be increased within the inflammatory infiltrate of the lamina propria and in the cytoplasm of the epithelial cells in patients with ulcerative colitis, whereas age-matched controls demonstrated no iNOS immunoreactivity. Localization of iNOS to the surface epithelial cells and areas of intense inflammation such as those near crypt abscesses has also been reported in Crohn's disease. In addition, increased iNOS expression and a higher percentage of iNOS-positive cells has been demonstrated in circulating monocytes of patients with IBD compared with healthy controls. Other studies have demonstrated iNOS expression within the colonic epithelium of normal mucosa. It appears that expression of the constitutive isoforms of NOS can also be altered in ulcerative colitis. nNOS expression was shown to be decreased in the muscularis mucosae of affected patients with ulcerative colitis, while eNOS was increased within the lamina propria of these patients. It is important to note that increased iNOS activity is not specific to IBD. When immunohistochemistry for iNOS was compared in patients with ulcerative colitis and infectious colitis, both populations demonstrated increased expression of iNOS in colonic epithelial cells, and the iNOS levels were reduced in the recovery phase of infectious colitis. The mechanism of NO-induced cellular damage is unknown, but one theory is that NO produced by iNOS interacts with oxygen radicals such as superoxide to produce peroxynitrite. This then reacts with tyrosine to form nitrotyrosine in cellular proteins. Singer et al. demonstrated nitrotyrosine staining in addition to intense expression of iNOS in the epithelium and adjacent lamina propria of the inflamed colonic mucosa of patients with ulcerative colitis and Crohn's disease. This pattern of nitrotyrosine immunohistochemistry was confirmed in a large sample of patients with ulcerative colitis from Japan.

3. Correlation of NO Production with Disease Activity in IBD

Measurement of mucosal NO production using NO electrodes in the rectums of patients with ulcerative colitis confirmed that increased iNOS activity and NO production correlated well with clinical and endoscopic indices of disease activity. A recent investigation confirmed that iNOS expression correlates with the acute and chronic inflammation score in pouchitis. Further, serum NO metabolites were increased in severe active ulcerative colitis. This increase correlated with C-reactive protein measurements, and both markers were found to decrease rapidly with glucocorticoid treatment in patients who were steroid responsive. In addition, serum nitrate levels in patients with active colitis have been reported to be directly proportional to erythrocyte sedimentation rate, leukocyte counts, and platelet counts. Elevated urinary nitrite levels have also been found to correlate with C-reactive protein levels in active ulcerative colitis. Exhaled NO has been measured using chemiluminescence in patients with ulcerative colitis, patients with Crohn's disease, and controls. Exhaled NO was significantly increased in both IBD groups relative to controls and correlated well with both the Crohn's disease activity index and the ulcerative colitis activity index. Interestingly, in this study, spirometric data in patients with Crohn's disease showed that increasing Crohn's disease activity index was associated with a worsening of FEV1 and FVC. The correlation of iNOS expression with disease activity in Crohn's disease is less obvious. A study from France demonstrated that iNOS activity was increased relative to controls in both Crohn's disease and ulcerative colitis, but that iNOS activity levels did not correlate with disease severity in the patients with Crohn's disease. A correlation of iNOS levels with disease activity did exist in ulcerative colitis patients, but in this study, iNOS activity levels were actually lower in the patients with ulcerative colitis versus the patients with Crohn's disease. Similarly, in a report by Kimura et al., iNOS activity levels correlated with the severity of inflammation in involved areas in patients with ulcerative colitis, but there was no difference in iNOS activity between normal and inflamed mucosa in patients with Crohn's disease.

4. NO in IBD: Animal Studies

The potential role of NO as a proinflammatory mediator in IBD has fueled numerous investigations in animal models of colitis. In addition to NO, peroxynitrite formed by the interaction of NO with reactive oxygen species that are also generated during inflammation is thought to be an important mediator of tissue damage in colitis. Initial studies implicated both NO and peroxynitrite in the pathophysiology of experimental IBD by demonstrating increased levels of iNOS and of nitrotyrosine staining. Additionally, when rats were given intrarectal peroxynitrite, they rapidly developed gross and histologic evidence of colitis. In rhesus macaques with idiopathic colitis, serum and urinary nitrate levels were increased, iNOS gene expression was increased, and iNOS activity was markedly elevated compared with controls. A wide array of studies have been conducted using selective and nonselective inhibitors of iNOS and iNOS-deficient mice to address the role of iNOS-derived NO in experimental IBD.

5. Beneficial Effect of iNOS

Despite the compelling evidence from the previous studies suggesting that iNOS is deleterious in acute colitis models and that selective inhibition can attenuate these effects, conflicting data are also available. Hosoi et al. showed that nonspecific inhibition with LNMMA worsened inflammation in a TNBS rat model of colitis. Ribbons et al. studied a rhesus monkey model that is prone to spontaneous colitis similar to human ulcerative colitis. Monkeys treated with oral L-NIL or AG had no improvement in diarrhea or histologic inflammatory scores. AG pretreatment has also been shown to worsen biochemical and macroscopic colitis scores in TNBS-treated rats. In the rat DSS colitis model, L-NAME and AG were both shown to worsen colitis, whereas in a rat TNBS colitis model, neither drug had an effect on colonic inflammation. Although these data suggested that NO may be protective in this model of colitis, when the same DSS-treated rats were given an NO donor, the colitis was also exacerbated. A follow-up study specifically addressed this concern by administering the NO donor diethylamine NONOate (DETA/NO) to mice with DSS-induced colitis. DETA/NO decreased loss of weight, increased colon length, and decreased production of interleukin (IL)-12 and IFN-γ, all suggesting a beneficial effect. Fiorucci et al. demonstrated that treatment of mice with NCX-1015, a nitric oxide derivative of prednisolone, protects mice from colitis induced by TNBS.

In the acetic acid model, iNOS−/− mice had a greater increase in acute macroscopic damage versus wild-type mice and had persistent colitis while the wild-type mice recovered by day 7. A similar exacerbation of acute colitis has also been reported in the TNBS model. However, in IL-10–deficient mice that spontaneously develop intestinal inflammation, deletion of the iNOS gene was not protective. Additionally, in the DSS model, iNOS-deficient mice were reported to be protected from the moderate colitis seen in wild-type mice. A second study in the DSS mouse model confirmed that iNOS deletion attenuated colitis and that treatment of wild type mice with 1400W was also protective. Thus, it is difficult to form a conclusion about the effect of iNOS deletion, based on the varying evidence in the above studies.

6. The Role of NO in IBD

The current data suggest that iNOS expression and activity is up-regulated in colon and intestinal tissues from patients with IBD and in experimental models of IBD. Studies with inhibitors in the latter do not lead to a clear conclusion since there are studies with both selective and nonselective iNOS inhibitors that show amelioration or worsening of disease. Similarly, in iNOS-deficient mice, both worsening and improvement of experimental IBD have been described. To attempt to address these discrepancies, it is important to consider that there are potential differences between species, mouse strains, models, and execution of the same models. Additionally, the time points considered may be of critical importance in influencing the results. However, it is also important to understand that within the same tissues, the role of NO may differ, depending on the cell type and the local concentration. As an example, iNOS activity is associated with cytotoxicity and apoptosis of gastrointestinal epithelial cells, while we have shown a protective effect of iNOS in human intestinal microvascular endothelial cells (HIMECs). Specifically, in collaborative studies with our laboratory spearheaded by Dr. David Binion at the Medical College of Wisconsin, we have reported that HIMECs express iNOS and produce micromolar quantities of NO in response to LPS or cytokines and that this results in inhibition of leukocyte adhesion in the activated endothelial cells. We then showed that iNOS expression and activity in HIMEC lines derived from regions of active inflammation from ulcerative colitis or Crohn's disease patients actually had deficient iNOS expression and NO production in response to activation, which correlated with hyperadhesion of leukocytes. Finally, cyclosporin A was found to increase leukocyte binding to activated HIMEC by blocking iNOS expression, which may provide insight into how patients can become refractory to the benefits of cyclosporin A in IBD. Thus, in these in vitro studies, an antiinflammatory role for iNOS-derived NO has been demonstrated, in that adhesion of leukocytes to activated HIMEC, a critical step in the inflammatory response, is attenuated by NO, and failure of this effect may occur in IBD.

8. NO in Hepatocellular Function

NO and Hepatocellular Protein Synthesis Hepatic dysfunction in sepsis. Patients with multisystem failure due to sepsis often have evidence of hepatocellular dysfunction manifested as elevations of bilirubin and liver-related enzymes as well as decreased serum albumin. The hypoalbuminemia probably has multiple mechanisms, but is at least partly due to a decrease in hepatocellular synthetic function. The mechanism of the hepatocellular dysfunction is poorly understood, but one hypothesis is that bacterial products such as endotoxins cause hepatocellular dysfunction directly or through the action of mediators released by macrophages such as interleukin 1 (IL-1) and tumor necrosing factor (TNF). Kupffer cells activated by endotoxin or cytokines are known to release multiple biologically active products such as IL-1 and TNF. These products of endotoxin-triggered Kupffer cells may mediate the decrease in hepatocyte protein synthesis in sepsis. Recent evidence indicates that NO may also mediate decreases in hepatocyte protein synthesis. NO and regulation of hepatocyte protein synthesis. Evidence supporting a role for NO as an inhibitor of hepatocyte protein synthesis in sepsis comes from a model that measures protein synthesis in rat hepatocytes that are cultured alone or in the presence of Kupffer cells. Coculture of macrophages or Kupffer cells with hepatocytes normally promotes hepatocyte protein synthesis. The addition of lipopolysaccharide endotoxin or killed Escherichia coli to cocultured hepatocytes and Kupffer cells but not to hepatocytes alone induces a profound decrease in hepatocyte protein synthesis without inducing hepatocellular death. The Kupffer cellmediated inhibition of hepatocyte protein synthesis is associated with production of nitrates, nitrites, and citrulline and is blocked by L-arginine analogs which competitively inhibit nitric oxide synthesis, suggesting that production of NO from L-arginine is required. In fact, direct chemical measurement of NO in the supernatant of endotoxin-stimulated hepatocyte Kupffer cell cocultures shows that NO is released. The time course of NO release parallels the decrease in hepatocyte protein synthesis. Because exogenous authentic NO also inhibits hepatocyte protein synthesis, these experiments indicate that NO is a mediator of decreased hepatocyte protein synthesis. Although stimulated Kupffer cells are the source for some of the NO production in this model, it also appears that humoral Kupffer cell products induce the formation of NO by hepatocytes and that hepatocyte production of NO also mediates decreased hepatocyte protein synthesis. When hepatocytes in culture are exposed to a cell-free supernatant from stimulated macrophages, hepatocyte protein synthesis is inhibited and the hepatocytes produce large quantities of NO. The inhibition of protein synthesis directly parallels NO production, and both are blocked by an inhibitor of NO synthesis, indicating that hepatocyte NO production mediates inhibition of hepatocyte protein synthesis. The production of NO in both Kupffer cells and hepatocytes appears to involve an inducible NO synthase as there is a delay of several hours between stimulation and the production of N0. Induction of NO synthase and NO production can be stimulated by endotoxin alone in Kupffer cells, but not hepatocytes. In duction of the hepatocyte NO synthase and the subsequent decrease in protein synthesis appear to require multiple cytokines produced by activated Kupffer cells, including TNF and IL-1. The mechanism by which NO inhibits hepatocyte protein synthesis is not known. It does not involve hepatocyte cell death because the effect is reversible, no hepatocellular enzymes are released, and the cells remain viable based on trypan blue exclusion. Although NO induces a rapid increase in cGMP levels in hepatocytes, a cGMP analog did not reproduce the inhibition of total protein synthesis, indicating that NO does not inhibit hepatocyte protein synthesis solely through the activation of guanylate cyclase. NO generated by macrophages inactivates complex I and complex II of the electron transport chain and the Krebs cycle enzyme aconitase. NO also inhibits these mitochondrial enzymes in hepatocytes, but the inhibition of mitochondrial respiration recovers within l-2 hours. Thus, it does not appear to explain the inhibition of protein synthesis, which persists for up to 12 hours. Northern blot analysis of hepatocyte albumin mRNA levels shows that NO has no effect on the relative abundance of hybridizable albumin mRNA, indicating that NO inhibits hepatocyte protein synthesis through an undefined translational or posttranslational mechanism. In platelets, NO can modulate the activity of an enzyme which may be involved in posttranslational steps of protein synthesis, raising the possibility that similar NO-sensitive enzymes may mediate the effects of NO on hepatocyte protein synthesis. In contrast, other work has shown that the inhibition of hepatocyte albumin synthesis by IL-1 and TNF was due to a pretranslational mechanism. The mechanism and role of NO in the modulation of hepatocyte protein synthesis during sepsis remain unclear. Physiological significance and speculation on clinical implications. The role of NO in hepatocellular dysfunction seen in humans with sepsis and endotoxemia is not known, and its potential therapeutic importance is unclear, because hepatocellular dysfunction and decreased albumin synthesis are not major determinants of survival in patients with multiorgan failure due to sepsis. However, it can be hypothesized that NO formation is involved in the decreased albumin synthesis seen in some patients with cirrhosis or in the impaired synthesis of coagulation factors found in other liver diseases. It is also possible that endogenous NO may modulate other hepatocellular functions such as the metabolism of glucose or lipids, although in the rat liver NO does not appear to mediate agonist-stimulated glycogenolysis. If NO is shown to have a role in human hepatocellular dysfunction, therapeutic interventions involving alterations in NO production may be possible.

9. NO in Cirrhosis

NO and the Hyperdynamic Circulation of Cirrhosis Splanchnic and systemic hemodynamics in cirrhosis. Portal hypertension is associated with hemodynamic disturbances in both the splanchnic and systemic circulation. Chronic portal hypertension is associated with splanchnic hyperemia, which may be important in maintaining chronic portal hypertension and may contribute to the development of the abnormalities of systemic hemodynamics seen in cirrhosis. These alterations in systemic hemodynamics include increased cardiac output and heart rate and decreased systemic vascular resistance and blood pressure. This hyperdynamic circulatory state develops before the onset of ascites and is found in 30%~50% of patients with cirrhosisl and in all animal models of portal hypertension. In patients with cirrhosis and ascites, the severity of the hemodynamic derangements is an independent predicator of survival, suggesting that the homeostatic responses to a decreased effective plasma volume may be harmful. The hyperdynamic circulatory state may be a reflection of a decrease in the "effective" plasma volume related to vasodilatation, which increases the vascular holding capacity. Decreased effective plasma volume may also play a key role in the development of important complications of cirrhosis including ascites, edema, and hepatorenal syndrome. According to the "underfill theory," the primary event in the development of these complications is a decrease in effective plasma volume related to primary peripheral vasodilatation. This leads to activation of volume receptors that activate the renin-angiotensin system, increase sympathetic outflow, and may increase release of atrial natriuretic factor and alter activity of the kallikrein-kinin system. These mediators increase renal sodium retention leading to ascites and edema and may alter renal blood flow leading to decreased glomerular filtration and the hepatorenal syndrome. This theory is supported by evidence from studies of humans with cirrhosis. Evidence suggests that systemic and splanchnic vasodilatation seen in cirrhosis with portal hypertension is due to decreased sensitivity to endogenous vasoconstrictors, and/or increased activity of endogenous vasodilators. The list of putative vasodilators in cirrhosis is long but no convincing evidence indicates a predominant role for any one of them. NO, endotoxin, and vasodilatation in cirrhosis. Recently it was proposed that NO acts as an endogenous vasodilator mediating the hyperdynamic circulation of cirrhosis. Endothelium-derived NO relaxes vascular smooth muscle and dilates arteries and veins contributing to the regulation of vascular tone in animals and humans. In animals, release of endothelium-derived NO triggered by acetylcholine or other agonists leads to peripheral vasodilatation, hypotension, and increased heart rate. If a substance that triggers endothelial NO formation is released into the circulation in cirrhosis, the resulting vasodilatation may produce the hyperdynamic circulatory state. There is evidence that supports a role for NO in the hyperdynamic circulation of cirrhosis and suggests that bacterial lipopolysaccharide endotoxin may be the trigger for increased endothelial NO formation. When vascular tissue is exposed to endotoxin or cytokines in vitro, NO synthase is induced in the endothelium and smooth muscle, and increased NO production leads to vasorelaxation and decreased responsiveness to vasoconstrictors. In the dog and rat, in vivo studies show that the vasodilatation and decreased vascular responsiveness that occur in response to endotoxin or cytokines is mediated by NO synthesis. The induction of NO synthase with increased production of NO may also explain the peripheral vasodilatation and decreased responsiveness to vasoconstrictors that occur in response to endotoxin infusion in humans. In humans who received a small bolus infusion of endotoxin, the peripheral vasodilatation was not immediate but gradually appeared during the 2 hours after infusion of the endotoxin and persisted for another 4-5 hours. These findings are consistent with the hypothesis that an inducible NO synthase is responsible for peripheral vasodilatation following endotoxin infusion in humans, because the lag time for NO production by the inducible NO synthase in animals is also 2 hours.

Recently, inhibitors of NO synthase were shown to increase blood pressure in two patients with septic shock, providing clinical support that NO production mediates endotoxin-induced hypotension. High levels of endotoxin are detected in the systemic circulation of a large percentage of cirrhotic patients. Thus, endotoxin could serve as a trigger for the induction of vascular NO synthase in cirrhosis. This systemic endotoxemia occurs in cirrhotic patients even in the absence of obvious infection and is thought to result from abnormal portal- systemic shunting, which allows systemic spillover of gut-derived bacterial endotoxin that would normally be cleared by the liver. Besides hemodynamic similarities between the effects of endotoxin infusion and the hyperdynamic circulation of cirrhosis, a number of lines of evidence support the hypothesis that endotoxin-triggered production of vascular NO causes the hyperdynamic circulation of cirrhosis. In a parabiotic model of portal hypertension, hyperdynamic circulation was found in the portal hypertensive animal, but not the normal parabiotic partner. These results suggest that the hyperdynamic circulatory state is not mediated by a transferable humoral factor, although they are consistent with the production of the hyperdynamic state by a locally acting mediator with an ultrashort half-life, such as NO. Because NO-induced relaxation of vascular smooth muscle involves an increase in cGMP production, increased production of cGMP would be expected in cirrhosis if NO mediates the hyperdynamic circulation. In the rat, urinary excretion of cGMP is a biologic marker of the level of vascular NO synthesis. In cirrhotic humans, significantly elevated urinary cGMP levels were detected even before the onset of ascites.lg6 Although atrial natriuretic peptide can increase cGMP production and may be elevated in cirrhosis, many of the cirrhotic patients with elevated cGMP did not have elevated levels of atrial natriuretic peptide, suggesting that increased urinary cGMP may have been related to increased vascular production and action of NO. In animals, glucocorticoids prevent the induction of vascular NO synthase by endotoxin and prevent the consequent decrease in vascular tone. Information on the hemodynamic effects of glucocorticoids in human cirrhosis is limited, but they may promote a diuresis in some patients with cirrhosis. Glucocorticoid therapy may decrease mortality in selected patients, but this is thought to be due to an effect on the activity of the underlying liver disease, and the precise effect of glucocorticoids on hemodynamics in humans with cirrhosis is unknown. In rats with chronic portal hypertension induced by portal vein ligation, a hyperdynamic circulatory state develops manifested by hypotension, systemic and splanchnic vasodilatation, and increased cardiac output. In this model, inhibition of NO synthesis, after induction of portal hypertension and the development of the hyperdynamic circulatory state, causes significant systemic and splanchnic vasoconstriction. There is also evidence that NO mediates arterial hypotension in rats with portal hypertension due to a carbon tetrachloride induced cirrhosis. These results support the hypothesis that endogenous NO synthesis mediates the systemic and splanchnic vasodilatation observed in portal hypertensive animals. A recent case report described an elevation in blood pressure in response to methylene blue infusion in a patient with liver failure due to alcoholic cirrhosis and severe hypotension resistant to pressor agents Methylene blue blocks the action of NO through the inhibition of guanylate cyclase, providing clinical evidence that the generation of NO is responsible for the hypotension of liver failure. Although the data described above support the hypothesis that endotoxin-mediated induction of vascular NO synthase leads to the hyperdynamic circulation of cirrhosis, other experimental results do not support this hypothesis. Although chronic endotoxin infusion in rats produced ascites, it did not change blood pressure, and the production of ascites appeared to be related to increased vascular permeability. In portal vein-ligated rats with a well-developed hyperdynamic circulatory state, no evidence of endotoxemia was found, production of tolerance to endotoxin by repeated exposure did not ameliorate the hyperdynamic state, and a reduction of intestinal bacteria with antibiotic therapy did not alter the splanchnic hemodynamic status. Speculation on clinical implications. If the endotoxemia of cirrhosis does lead to induction of vascular NO synthase, and if increased NO synthesis and release is responsible for the hyperdynamic circulation of cirrhosis, a number of potential therapeutic interventions are suggested. Blockade of this pathway might prevent or reverse some of the complications of cirrhosis that may be related to systemic vasodilatation, including ascites, edema, and the hepatorenal syndrome. Selective intestinal decontamination with quinoline antibiotics might effectively reduce the aerobic gram negative bowel flora, decrease endotoxemia, and prevent the induction of vascular NO synthase. Glucocorticoids or, preferably, agents with greater specificity and less toxicity might be used to prevent induction of NO synthase by endotoxin and prevent the development of the hyperdynamic circulatory state. Inhibitors of NO synthase, used as medications to stop the increased vascular production of NO in the hyperdynamic circulatory state, could prove to be effective therapy for cirrhotic ascites or hepatorenal syndrome. NO and Treatment of Portal Hypertension Medical therapy to prevent variceal bleeding. Hemorrhage from esophageal varices in patients with portal hypertension is associated with high morbidity and mortality despite aggressive resuscitation and therapy. Because of this, attempts have been made to prevent variceal hemorrhage by lowering portal venous pressures in patients with cirrhosis. Beta-blockers may prevent variceal bleeding, but the long-term efficacy of this therapy remains unclear, and they have the potential for adverse effects. Preliminary evidence suggests that pharmacological modulation of NO activity may be a useful new mechanism to lower portal pressures. Organic nitrates cause vasodilatation by activating guanylate cyclase in vascular smooth muscle. The principal mode of activation of guanylate cyclase by these drugs may be through the liberation of NO from the organic nitrate. Organic nitrates reduce portal pressure, and their use in the treatment of cirrhotic patients has been advocated. However, organic nitrates also reduce systemic blood pressure, and chronic administration can induce tolerance, which would limit their effectiveness in the chronic therapy of portal hypertension. Molsidomine, another drug that acts through the generation of NO, has been investigated as therapy for portal hypertension. Molsidomine is one of a class of drugs called syndnonimines and is a longacting vasodilator that is an established antianginal agent. It is metabolized to form an active metabolite called SIN-1, which produces vasodilatation by acting as an NO donor, leading to the stimulation of cGMP production in vascular smooth muscle. Unlike the organic nitrates, molsidomine has little or no effect on systemic blood pressure in patients with normal livers, and it does not induce tolerance. In rats with carbon tetrachloride-induced cirrhosis and portal hypertension, molsidomine causes a significant decrease in portal venous pressure, apparently by reducing portal-collateral resistances without affecting hepatic blood flow. However, molsidomine also reduces systemic arterial pressure, increases portal venous inflow, and may increase collateral blood flow. Although a decrease in portal venous pressure would be expected, to be beneficial in patients with cirrhosis, the other effects seen with molsidomine in this animal study could be detrimental by decreasing renal perfusion or increasing pressure in collateral veins such as esophageal varices. In patients with cirrhosis and portal hypertension, a single dose of molsidomine causes a significant and sustained decrease in portal venous pressure. Molsidomine also causes moderate decreases in he-patic blood flow and systemic arterial pressure, but intrinsic hepatic clearance of indocyanine green was not reduced, indicating that molsidomine reduced portal pressure without impairing liver elimination function. These results suggest that molsidomine may be an effective and safe long-term treatment for portal hypertension, but its efficacy in preventing variceal bleeding will need to be determined.

REFERENCES

[1] Guslandi, M. Nitric oxide: an ubiquitous actor in the gastrointestinal tract. *Dig Dis.*, 1994,12,28-36.

[2] Park, JH; Kwon, JG; Kim, SJ; et al. Alterations of colonic contractility in an interleukin-10 knockout mouse model of inflammatory bowel disease. *J Neurogastroenterol Motil.*, 2015, 21, 51-61.

[3] Dijkstra, G; van Goor, H; Jansen, PL; et al. Targeting nitric oxide in the gastrointestinal tract. *Curr Opin Investig Drugs.*, 2004, 5, 529-536.

[4] Izzo, AA; Mascolo, N; Capasso, F. Nitric oxide as a modulator of intestinal water and electrolyte transport. *Dig Dis Sci.*, 1998, 43, 1605-1620.

[5] Shah, V ; Lyford, G; Gores, G; et al. Nitric oxide in gastrointestinal health and disease. *Gastroenterology.*, 2004, 126, 903-913.

[6] Prasad, P. The role of nitric oxide in gastrointestinal health and disease. *Gastroenterology.*, 2004, 127, 1014.

[7] Vahedian, M; Nabavizadeh, F; Vahedian, J; et al. Lead exposure changes gastric motility in rats: role of nitric oxide (NO). *Arch Iran Med.*, 2011, 14, 266-269.

[8] Pullen, NA; Fillmore, HL. Induction of matrix metalloproteinase-1 and glioma cell motility by nitric oxide. *J Neurooncol.*, 2010, 96, 201-209.

[9] Lahiri, M; Martin, JH. Nitric oxide decreases motility and increases adhesion in human breast cancer cells. *Oncol Rep.*, 2009, 21, 275-281.

[10] Hassanpour, H; Mirshokrai, P; Shirazi, A; et al. Effect of nitric oxide on ram sperm motility in vitro. *Pak J Biol Sci.*, 2007, 10, 2374-2378.

[11] Bult, H; Boeckxstaens, GE; Pelckmans, PA; et al. Nitric oxide as an inhibitory non-adrenergic non-cholinergic neurotransmitter. *Nature.*, 1990, 345, 346-347.

[12] Martin, MJ; Jimenez, MD; Motilva, V. New issues about nitric oxide and its effects on the gastrointestinal tract. *Curr Pharm Des.*, 2001, 7, 881-908.

[13] Wallace, JL; Miller, MJ. Nitric oxide in mucosal defense: a little goes a long way. *Gastroenterology.*, 2000, 119, 512-520.

[14] Kubes, P; Wallace, JL. Nitric oxide as a mediator of gastrointestinal mucosal injury?-Say it ain't so. *Mediators Inflamm.*, 1995, 4, 397-405.

[15] Huang, Y; Lemberg, DA; Day, AS; et al. Markers of Inflammation in the Breath in Paediatric Inflammatory Bowel Disease. *J Pediatr Gastroenterol Nutr.*, 2014.

[16] Slomiany, BL; Slomiany, A. Involvement of p38 MAPK-dependent activator protein (AP-1) activation in modulation of gastric mucosal inflammatory responses to Helicobacter pylori by ghrelin. *Inflammopharmacology*., 2013, 21, 67-78.

[17] ten Oever, J; Mandon, J; Netea, MG; et al. Pulmonary infection, and not systemic inflammation, accounts for increased concentrations of exhaled nitric oxide in patients with septic shock. *J Breath Res*., 2013, 7, 036003.

[18] Zanini, GM; Cabrales, P; Barkho, W; et al. Exogenous nitric oxide decreases brain vascular inflammation, leakage and venular resistance during Plasmodium berghei ANKA infection in mice. *J Neuroinflammation*., 2011, 8, 66.

[19] Lee, JS; Paek, NS; Kwon, OS; et al. Anti-inflammatory actions of probiotics through activating suppressor of cytokine signaling (SOCS) expression and signaling in Helicobacter pylori infection: a novel mechanism. *J Gastroenterol Hepatol*., 2010, 25, 194-202.

[20] Campisi, J; Leem, TH; Fleshner, M. Acute stress decreases inflammation at the site of infection. A role for nitric oxide. *Physiol Behav*., 2002, 77, 291-299.

[21] Maeda, H; Akaike, T. Nitric oxide and oxygen radicals in infection, inflammation, and cancer. *Biochemistry* (Mosc)., 1998, 63, 854-865.

[22] Rose, JW; Hill, KE; Wada, Y; et al. Nitric oxide synthase inhibitor, aminoguanidine, reduces inflammation and demyelination produced by Theiler's virus infection. *J Neuroimmunol*., 1998, 81, 82-89.

In: Nitric Oxide
Editor: Dominic M. Evans
ISBN: 978-1-63482-961-8

Chapter 2

S-Nitrosylation: A Double-Edged Sword in the Progression and Inhibition of Cancer

Kimberly Hallowell*[1, 2]*, Laura St. Clair*[2]*,
***Dana Jarigese*[1] *and Mark A. Brown*[3,4,*]**

[1]Department of Animal Science, Colorado State University, Fort Collins, CO, US
[2]Department of Biology, Colorado State University, Fort Collins, CO, US
[3]Department of Clinical Sciences, Colorado State University, Fort Collins, CO, US
[4]Epidemiology Section, Colorado School of Public Health, Fort Collins, CO, US

Abstract

This chapter examines the complex relationship between S-nitrosylation and cancer biology. As a double-edged sword, S-nitrosylation has been shown to be involved in malignant transformation, cell proliferation, induction of angiogenesis, and inhibition of DNA repair and apoptotic pathways. However, on the reverse edge, this post-translational modification has the ability to induce pro-apoptotic pathways and

* Corresponding Author: Mark A. Brown, Department of Clinical Sciences, Colorado State University, Fort Collins, CO 80523-1052, Email: M.Brown@colostate.edu.

upregulate key DNA repair genes. Herein, we discuss the conflicting signals of S-nitrosylation and its potential role in the clinical management of cancer.

INTRODUCTION

It has long been recognized that nitric oxide (NO) is heavily involved in cell signaling. NO is highly diffusible and short lived making it ideal for carrying highly specific messages across cellular membranes. This molecule is a reactive free radical which is produced by NO synthases (NOS) from the amino acid L-arginine [1]. There are three principal NOS isoforms, all of which synthesize NO under different cellular conditions. Neuronal NOS (nNOS) and endothelial NOS (eNOS) are calcium-dependent. nNOS and eNOS are responsible for the production of NO within the central nervous system (CNS) and outside the CNS, respectively. A third isoform, inducible NOS (iNOS) is calcium-independent and synthesizes NO in a rapid and highly concentrated manner in immunological signaling pathways [2, 3].

In animals, NO signaling can be categorized into two different mechanisms, namely, cGMP-dependent and cGMP-independent [3]. cGMP-dependent pathways are also known as "classical" and "less classical" NO signaling [2]. These pathways involve the activation of the hemoprotein, soluble guanylyl cyclase (sGC), which produces a second messenger cyclic GMP. This second messenger is responsible for regulating many cellular activities such as vascular dilation and neurotransmission by way of activation of ion channels and cGMP-governed kinases [2, 4]. cGMP-dependent pathways utilize NO synthesized by eNOS and nNOS [2]. The signaling pathways categorized as cGMP-independent involve the modification of other proteins by NO, either directly through covalent modifications, or indirectly through the modification of proteins upstream in the signaling pathway [5]. NO within the cGMP-independent pathways is produced by iNOS [2]. The best studied of these mechanisms is protein S-nitrosylation.

S-nitrosylation is a reversible post-translational modification (Figure 1) which occurs through the attachment of NO to a reactive thiol group in cysteine residues of proteins. This reaction results in the production of S-nitrosothiol (SNO) which is a fairly stable protein [1-4, 6-20]. An increasing number of studies illustrate the importance of S-nitrosylation based upon its impact on cellular activities. S-nitrosylationthiol selectivity is governed by protein-protein interactions and localization of NO in relation to the proteins.

The reverse process, protein denitrosylation, is governed by the same two processes and is important in maintaining and modifying the levels of SNO-proteins within the cell [19]. S-nitrosoglutathione (GSNO) reductase and thioredoxin are two enzymes which function prominently in the denistrosylation of proteins [1].

As mechanisms of post-translational modification, both S-nitrosylation and denitrosylation have key impacts on cellular function. These processes occur frequently and are a part of the normal activities of the cell. However, the dysregulation of both pathways has been linked to numerous human pathologies including neurodegenerative diseases, cardiovascular disorders, musculoskeletal diseases, and tumorigenesis [2, 4, 7]. Recent research has focused on the implications of S-nitrosylation in the development of these diseases and the potential for therapeutics targeting this mechanism. In this chapter we focus on the impacts of S-nitrosylation on cancer and the potential for new or combined therapies to be developed which target S-nitrosylation or denitrosylation.

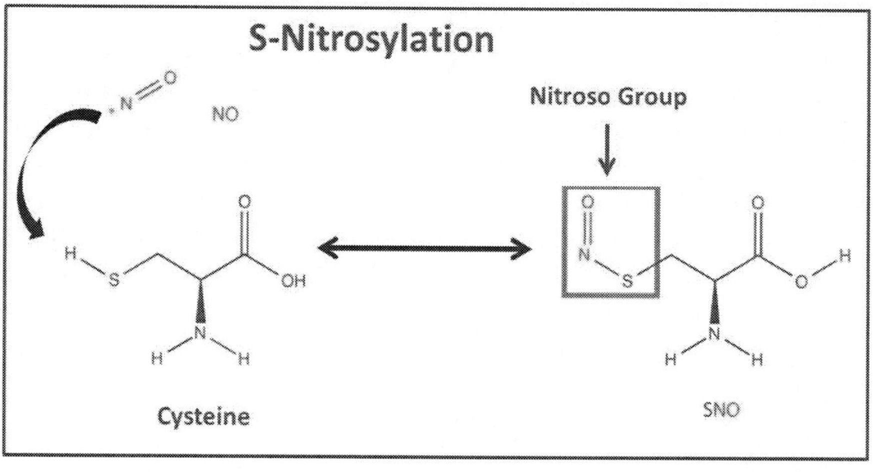

Figure 1. The Reversible S-nitrosylation Reaction.

S-NITROSYLATION AND CANCER PROGRESSION

Cancer is a multi-faceted disease in every sense, from its causes to its outcomes. Recent research has implicated S-nitrosylation as a causative agent in the breakdown of regulation in cellular processes that can push a cell toward a malignant state. In this section we examine the tumorigenic effects of S-nitrosylation and the proteins it influences. Some of the pathways in which S-nitrosylation has been identified play a role in malignant transformation, cell proliferation, inhibition of apoptosis, inhibition of DNA repair, and angiogenesis.

Malignant Transformation and Cell Proliferation

The beginnings of any cancer are manifest in malignant transformation and cell proliferation. This is the conversion of a normally functioning cell into an unhealthy and potentially dangerous state followed by rapid replication of the malfunctioning cell. A number of gene mutations have been identified that can induce malignant transformation and excessive proliferation. S-nitrosylation often serves to amplify the negative impacts of these mutated proteins.

C-Src

The tyrosine kinase c-Src is responsible for the phosphorylation of a number of proteins involved in metastasis, cell proliferation, and cell invasion [20]. It has been demonstrated that S-nitrosylation of c-Src increases its kinase activity, thus magnifying its downstream effects [3]. The catalytic domain of c-Src is closely linked to a C-terminal tail that contains numerous cysteine residues. It is at this regulatory C-terminus that S-nitrosylation of c-Src is believed to occur. By activating c-Src, S-nitrosylation contributes to the ability of cancerous cells to invade other tissue types and further proliferate. Over-activation of c-Src has been implicated in a number of different cancer types, including breast, colon, and prostate cancers, and may be an indicator of cancer aggressiveness and invasiveness [15].

PTEN

The phosphatase and tensin homolog (PTEN) appears to have tumor suppressor activity in nearly all cells in the human body and is one of the most commonly mutated tumor suppressors in human cancers. It serves primarily as a regulator of a number of different cellular pathways, including those associated with phosphatidylinositol-3-kinase (PI3K) and protein kinase B (AKT). PI3K produces a large number of inositol phospholipids, the primary substrates of PTEN. PTEN thus acts as an antagonist of PI3K signaling, inhibiting its downstream effects [20]. PI3K serves as an activator of AKT, a kinase that phosphorylates a number of proteins as a means of regulating pathways that are involved in metastasis, cell proliferation, and anti-apoptotic effects. The AKT pathway is one of the most commonly hyper-activated pathways in tumors [21]. S-nitrosylation of PTEN has an inhibitory effect on its phosphatase activity, thereby preventing it from antagonizing the PI3K pathway. This leads to unchecked activation of AKT, which ultimately induces over-stimulation of the pathways involved in malignant transformation and cell proliferation [22]. The PI3K/AKT pathways are also involved in HIF-1α stabilization, which can promote angiogenesis as is discussed below [18].

Inhibition of Apoptosis

Apoptosis refers to the highly regulated process of cell death which occurs normally in cells as a result of the activation of pro-apoptotic proteins called caspases. Some caspases, when nitrosylated, may inhibit apoptosis. One such caspase is caspase 3 which is involved in mitochondrial mediated cell death. When nitrosylated, it interacts with a Fas receptor which promotes its release into the cytoplasm where it can no longer impart its apoptotic effects [4]. Caspases 8 and 9 are the main apoptotic proteins governing the intrinsic and extrinsic pathways, respectively [12]. S-nitrosylation of proteins within these pathways induces conformational changes within the target protein which, in turn, affects the functionality of that protein. Inhibition of apoptosis in cells which have incurred DNA damage, chromosome damage, or cell cycle modification result in an enhanced risk of cancer [20].

P53

P53 is a key tumor suppressor which has often been called the "guardian of the genome" for its role in cellular response to DNA damage and genomic alteration. When p53 is activated, it arrests the cell cycle so that either DNA repair or apoptosis may occur, depending on the severity of the incurred damage [20]. Nitrosylated p53 undergoes a conformational change resulting in the inability to bind DNA and induce DNA repair or apoptosis [23]. This malfunction of p53 plays a major role in the development of a wide range of neoplasias.

Bcl-2

Bcl-2 is an anti-apoptotic protein which prevents apoptosis by inhibiting release of mitochondrial cytochrome c [3, 12, 20]. Bcl-2 is a key protein within the intrinsic pathway of apoptosis mediating mitochondria-dependent cell death. S-nitrosylation of Bcl-2 at one of two cysteine residues (C158 and C229) stabilizes the protein and prevents its degradation. Stabilization allows the overexpression of Bcl-2 within the cell which, in turn, promotes cell survival. In cells which exhibit irreparable damage, this over accumulation of Bcl-2 can lead to survival of cells which would typically be destroyed, thereby enhancing the potential for the development of tumorigenic cells [3, 12]. S-nitrosylated Bcl-2 is particularly problematic for cancer patients undergoing treatment with cytotoxic therapeutic agents, as Bcl-2 often inhibits the destruction of these targeted cells. Additionally, certain chemotherapeutic agents, such as Cisplatin, which specifically target Bcl-2 because of its anti-apoptotic properties, are unable to destroy the nitrosylated form of the molecule [3, 20].

FLIP

FLIP is an anti-apoptotic protein which regulates the extrinsic pathway of apoptotic signaling [12]. This extrinsic pathway is also known as the death receptor pathway, which acts through the interaction of a specific ligand and the death receptor on the extracellular surface of the cell. This interaction ultimately results in the activation of caspase-8 which stimulates apoptosis. FLIP disrupts this pathway by inhibiting the activation of caspase-8 when it is recruited to the complex formed on the inner surface of the membrane near the death receptor. Production and degradation of FLIP are key regulatory processes in the apoptotic pathway. While degradation of FLIP increases the likelihood that apoptosis will occur, overexpression of FLIP can cause increased resistance to apoptosis. S-nitrosylation of FLIP diminishes its normal degradation thereby leading to FLIP-mediated resistance to apoptosis [12].

MKP-1

Mitogen-activated protein kinase (MAPK) phosphatase -1 (MKP-1) inactivates MAPKs through dephosphorylation. A main feature of MKPs is a catalytic cysteine, Cys258 in MKP-1. MKP-1 levels are elevated in many tumors and are associated with increased resistance to chemo- or radio-therapy [19-20]. When NO is present, S-nitrosylation of MKP-1 occurs at this catalytic Cys258 and promotes protein stability and phosphatase activity [26]. In head and neck cancers, S-nitrosylated MKP-1 is known to enhance anti-apoptotic factors of tumor cells following gamma irradiation [9].

Inhibition of DNA Repair

DNA mutation in eukaryotic cells, as a result of exposure to carcinogens and radiation or miscopies during the DNA replication process, is not uncommon. This damage, if not repaired, can be detrimental, disrupting the function of entire organ systems. In certain genes, DNA damage may ultimately lead to tumorigenesis. Eukaryotic cells have a wide variety of measures in place to prevent these mutations from proliferating into the next generation, many of which include key proteins which proofread and correct mistakes that are made. These proteins, like others, are subject to post-translational modification, including S-nitrosylation [19].

AGT

O^6-alkylguanine-DNA-alkytransferase (AGT) is a key protein in direct DNA repair. AGT repairs highly mutagenic and cytotoxic O^6- alkylguanines by transferring an alkyl group from guanine to a cysteine in the active site of the enzyme. This reaction leads to the irreversible inactivation of AGT and restores the guanine to its proper structure. These O^6-alkylguanines are mismatched by DNA polymerase to thymine which may result in GC to AT mutations, DNA double strand breaks, and cell death [19]. AGT plays a critical role in preventing tumorigenesis by preventing O^6-alkylguanines from becoming incorporated into the DNA strand [3]. S-nitrosylation of AGT occurs at cysteine 145, which is the same cysteine located in the active site of the enzyme that would transfer the alkyl group. S-nitrosylation of AGT destabilizes the protein, decreasing its half-life to less than a few hours, thereby resulting in its rapid degradation. This reaction is irreversible and prevents routine DNA repair. Loss of AGT within the cell increases genetic susceptibility to mutation and increases the risk of tumorigenesis [3, 19-20].

8-Oxoguanine Glycosylase (Ogg1)

Base excision repair (BER) is a critical process in DNA repair. The BER pathway repairs single base modifications, such as those involved in DNA alkylation or oxidation. Repair by the BER pathway prevents mutagenesis and tumorigenesis. 8-oxoguanine glycosylase (Ogg1) is a glycosylase which plays a critical role in the BER pathway. Ogg1 exhibits specificity for 8-oxoguanine (8-oxoG) [20] which is an oxidative DNA lesion which increases the likelihood for GC to TA transversions. When Ogg1 is active and properly functioning, it is effective at repairing such oxidative stress. Ogg1 has 2 DNA-binding motifs, a helix-hairpin-helix motif and a Cys-His-type zinc finger motif. The presence of cysteine residues makes the zinc finger motif susceptible to S-nitrosylation. When nitrosylated the catalytic activity of Ogg1 is disturbed, thereby preventing it from removing oxidative damage to DNA. S-nitrosylation results in irreversible loss of catalytic activity of Ogg1 due to the disruption in its tertiary structure. Without the removal of oxidative DNA damage, there is an increased risk of mutagenesis and carcinogenesis within the cell [24].

APE1

Apurinic-apyrimidinic endonuclease 1 (APE1) is another protein critical in the BER pathway. APE1 normally resides within the nucleus where it aids in DNA repair as well as regulating transcription. S-nitrosylation of APE1 results in its translocation from the nucleus to the cytoplasm where it can no longer perform activities of the BER pathway. Localization of APE1 within the cytoplasm has been discovered in colon adenomas, breast cancer, and hepatocellular carcinomas, among others [18].

Angiogenesis

Angiogenesis plays a critical role in the maintenance and progression of tumors. One of the key steps in the growth of any tumor is offsetting the balance between pro- and anti-angiogenic processes to allow formation of new blood vessels. Not only do new blood vessels allow for transport of nutrients into and waste out of the tumor cells, they also facilitate the process of metastasis [4].

MKP-7

Mitogen-activated protein kinase (MAPK) phosphatase-7 (MKP-7) is from a family of proteins that, under normal circumstances, serve as negative regulators of MAPK function. MKP-7 works to inactivate the p38 and JNK MAP kinases produced in response to stress stimuli which, ultimately, halts the migration of endothelial cells during angiogenesis [25]. S-nitrosylation of MKP-7 reduces its phosphatase activity by altering a key cysteine residue within the active site of the enzyme, preventing it from successfully dephosphorylating JNK [20]. This leads to extended stability of JNK, promoting endothelial cell migration and, thereby, enhancing the angiogenic process [3].

HIF-1α

Hypoxia-inducible factor 1α (HIF-1α) is a pro-angiogenic transcription factor that is typically expressed only in hypoxic conditions to promote blood vessel formation leading to normalization of O_{2-} levels [26]. HIF-1α serves as a key activator of pathways involved in angiogenesis, cell proliferation, and cell survival [13]. Under normal circumstances, HIF-1α is degraded when bound by pVHL, the product of a tumor suppressor gene. Nitrosative stress, however, leads to S-nitrosylation of a key residue in HIF-1α's degradation domain, increasing its stability even under normoxic conditions [20]. This regulatory failure leads to unchecked activation of angiogenic and proliferative pathways mediated, in part, by HIF-1α. This, in turn, results in increased radiotherapy resistance in cancer cells and promotion of angiogenic pathways. As discussed above, deregulation of HIF-1α also appears to be influenced by loss of function in tumor suppressor genes such as PTEN and p53 [4, 13].

Dynamin

The dynamin family of GTPases is largely involved in endocytotic and secretory pathways. In relation to cancer and cell survival, dynamin proteins influence endocytosis of receptor proteins which increases survival signaling, particularly in endothelial cells. By increasing endothelial cell survival and growth, dynamin has the ability to promote the maintenance and growth of the body's vasculature [27]. S-nitrosylation of dynamin enhances its GTPase activity and increases general assembly of the protein. By upregulating GTPase activity, S-nitrosylation increases dynamin-mediated endocytosis of surface receptors and subsequent downstream cell survival signaling [20].

S-Nitrosylation and Cancer Inhibition

As S-nitrosylation is a double-edged sword, the tumor-slaying edge of the sword must not be forgotten. Although S-nitrosylation promotes cancer in numerous ways it has also been shown to help induce apoptosis in cancer cells and to block inhibitors of important DNA maintenance genes such as p53. These mechanisms illustrate the promise of S-nitrosylation as a potential factor in the clinical management of cancer.

Proteins of Cancer Inhibition

Fas

The Fas receptor is a member of the tumor necrosis factor (TNF) superfamily and is involved in apoptotic signaling. Under normal physiological conditions, the binding of the Fas ligand, FasL, to the receptor causes activation of a number of caspases, leading to the formation of a death-inducing signaling complex (DISC) [20]. Cancer cells typically have normal expression of Fas but the binding of FasL does not induce apoptotic signaling in these cells and the DISC, consequently, does not form. S-nitrosylation of Fas on a cytoplasmic cysteine residue was shown to increase its ability to recruit lipid rafts. This facilitates formation of the DISC complex and re-sensitization to the apoptotic signaling of Fas [18].

Hdm2

Hdm2, the human homolog of a protein first found on a mouse double minute chromosome (mdm2), is a known regulator of p53 activity in the cell [16]. Under normal circumstances, Hdm2 binds to p53 and represses its activity by blocking its transactivation domain. Hdm2 is also capable of blocking the transcription of p53 and is produced excessively in a number of different human cancer types. It has been shown that S-nitrosylation of Hdm2 can prevent it from binding to p53, allowing it to aid in DNA repair and the destruction of tumorigenic cells [16]. Increasing S-nitrosylation of Hdm2 could play a role in preventing malignant transformation of cells with functional p53 genes and thereby prevent a variety of cancers.

GAPDH

Glyceraldehyde-3-phosphate dehydrogenase (GAPDH) is involved in a number of different cellular processes and, following S-nitrosylation, can initiate apoptosis in cells [18]. Under normal circumstances, GAPDH functions in the glycolytic pathway. However, recent studies have emphasized its role in an apoptotic pathway when a cell is under stress [18]. When GAPDH recognizes cell stress, S-nitrosylation occurs at a catalytic cysteine residue on the protein, allowing it to bind to Siah1 and a ubiquitin ligase, forming a large complex. This complex translocates to the nucleus where it facilitates the degradation of substrates of Siah1. The products of this degradation induce a condition of cytotoxicity, ultimately leading to the death of the cell [10]. Mitochondrial GAPDH is also capable of inducing cell death in conjunction with several other key enzymes by halting cellular production of ATP, which occurs when high levels of NO are present due to cell stress [18]. Thus, NO-driven stress signals and S-nitrosylation of GAPDH can help prevent the proliferation of cancerous cells.

NF-κβ

Nuclear factor-κB is a transcription factor involved primarily in the inflammatory response, cell proliferation, and cell survival [13]. Under normal conditions, NF-κB actually plays an anti-apoptotic role by binding specific DNA sequences to activate anti-apoptotic proteins. This role is counteracted by inhibitory IκB proteins that complex with and sequester NF-κB in the cytoplasm to prevent it from accessing DNA. Activators of NF-κB phosphorylate IκB proteins to prevent them from forming a complex with NF-κB [13]. Regulation of NF-κB by S-nitrosylation is multifaceted. It acts directly via S-nitrosylation of an amino acid residue typically involved in NF-κB's DNA binding activities, preventing it from stimulating anti-apoptotic proteins [4]. It also acts indirectly by S-nitrosylating IκB, preventing activators of NF-κB from phosphorylating IκB. This increases IκB's ability to sequester and inhibit NF-κB's normal role, thereby leading to increased chances of apoptosis in tumor cells [14].

Using S-Nitrosylation to Increase the Susceptibility of Cancerous Cells to Treatment

S-nitrosylation affects different cells in varying ways. Nitrosylation of certain proteins in certain cells may lead to apoptosis while nitrosylation of others may inhibit the same process. Making the situation even more complex, S-nitrosylation of different cis residues on the same protein often imparts distinct outcomes. Moreover, the concentration and duration of exposure to NO also affect its impact on the cell [4]. Whether by inhibition or induction of apoptosis, S-nitrosylation exhibits a remarkable correlation with cancer progression and proliferation. Therefore, the process of S-nitrosylation presents a potential target for future cancer therapeutics.

Inhibition of S-Nitrosylation

With regard to proteins such as Bcl-2, FLIP, and MKP-1, S-nitrosylation inhibits apoptosis, which can contribute to tumor development. In these cases preventing the nitrosylation event can lead to decreased chance of tumor development or cancer progression. S-nitrosylation of the protein HIF-1α stabilizes the cell and leads to resistance of tumors to radiation. In this way inhibiting the nitrosylation of HIF-1α acts in partnership with radiotherapy to disrupt cancerous cells and reduce tumor growth [8].

S-Nitrosylation as a Potential Cancer Therapeutic

As previously discussed, there are many circumstances in which S-nitrosylation increases the chances of apoptosis. Increasing the chances of apoptosis in cancerous or pre-cancerous cells is clearly a target for a wide range of cancer treatment methods. If the probability of apoptosis is increased, the cell is more susceptible to outside pressures which may trigger cell death. NO is a reactive species which, in the cell, can induce mitochondrial-mediated apoptosis. NO may also increase oxygen delivery to tumors. The presence of oxygen enhances the effectiveness of radiotherapy by producing radical oxygen species. These reactive species interact with DNA damaged by radiation and prevent DNA repair mechanisms from functioning properly. When NO is introduced to tumor sites in conjunction with radiation, the probability that apoptosis occurs in those cells is increased [2]. This may lead to shorter bouts of less severe treatments for cancer patients. When S-nitrosylation induces apoptotic pathways, the tumor cells, in which the proteins are nitrosylated are sensitized to chemotherapy, radiation, and immunotherapy [6]. NOS-dependent apoptosis has been observed in thyroid carcinoma cells when treated with THN-related apoptosis-inducing ligand (TRAIL). This cancer therapeutic has been noted as tumor-selective and it is believed that these effects are derived from S-nitrosylation of a particular protein within the thyroid carcinoma cells [8]. If other tumor cells can be targeted as such, there is great potential for S-nitrosylation to provide a more selective cancer treatment, thereby reducing collateral damage to other cells in the body.

CONCLUSION

The signaling capacity of S-nitrosylation has generated a great deal of interest as an increasing number of targets of this post-translational modification have been discovered. Of particular importance is the growing list of S-nitrosylation protein substrates whose impacts on cancer biology are altered as a result of this modification. Under some conditions, S-nitrosylation exhibits the capacity to serve as a regulatory mechanism to constrain the progression of cancer. Under other conditions, as with most signal events, S-nitrosylation can be manipulated in tumorigenic pathways leading to the altered activity of proteins which are essential in the control of cell proliferation and survival. Whether involved in the progression or inhibition of cancer, an understanding of the signal capacities of this double-edged sword promises to pave the way to a formidable new approach for augmenting anti-cancer processes while curbing mis-regulated pathways.

Conflicts of Interest

The authors declare no conflicts of interest.

Acknowledgments

MAB thankfully acknowledges support from the National Science Foundation (1060548), the CVMBS College Council, and the Flint Animal Cancer Center.

References

[1] Kovacs, I., Lindermayr, C. (2013). Nitric oxide-based protein modification: formation and site-specificity of protein S-nitrosylation. *Frontiers in Plant Science* 4: 1-10.

[2] Kenote, N. J, Medeiros, T. N., Jarigese, D. M., Edwards, M. A., Brown, M. A. (2014) *Nitric oxideas an adjuvant therapeutic in the clinical management of cancer.* In Frontiers in Anti-cancer Drug Discovery, 3rd Edition. Atta-ur-Rahman and M. Iqbal Choudhary, Ed.: Bentham Science Publishers, Ltd. Sharjah, U.A.E.

[3] Medeiros, T. N., Jarigese, D. M., Edwards, M. A., Brown, M. A. (2015). *Impacts of S-Nitrosylation in Cancer.* In NO and Cancer: Pathogenesis and Therapy. Benjamin Bonavida, Ed.: Springer Publishers: New York, NY.

[4] Aranda, E., López-Pedrera, C., De La Haba-Rodriguez, J. R., Rodriguez-Ariza, A. (2012). Nitric oxide and cáncer: the emerging role of S-nitrosylation. *Curr Mol Med* 12(1):1875-5666.

[5] Muntané, J., De la Mata, M. (2010). Nitric oxide and cáncer. *World Journal of Hepatology* 2(9): 337-344.
[6] Anand, P., Stamler, J.S. (2012). Enzymatic mechanisms regulation protein S-nitrosylation: Implications in health and disease. *J Mol Med* 90 (3): 233-255.
[7] Feng, X., Sun, T., Bei, Y., Ding, S., Zheng, W., Lu, Y., Shen, P. (2013). S-nitrosylation of ERK inhibits ERK phosphorylation and induces apoptosis. *Scientific Reports* 3: 1-6.
[8] Foster, M. W., Hess, D. T., Stamler, J. S. (2009). Protein S-nitrosylation in health and disease: A current perspective. *Trends Mol Med* 15(9): 391-404.
[9] Guan, W., Sha, J., Chen, X., Xing, Y., Yan, J., Wang, Z. (2012). S-nitrosylation of mitogen activated protein kinase phosphatase-1 supresses radiation-induced apoptosis. *Cancer Lett* 314(2): 137-146.
[10] Hara, M. R., Cascio, M. B., Sawa, A. (2006). GAPDH as a sensor of NO stress. *Biochim Biophys Acta* 1762: 502-509.
[11] Hess, D. T., Stamler, J. S. (2012). Regulation by S-nitrosylation of protein post-translational modification. *J Biol Chem* 287(7): 4411-4418.
[12] Iyer, A. K. V., Azad, N., Wang, L., Rojanasakul, Y. (2008). Role of S-nitrosylation in apoptosis resistance and carcinogenesis. *Nato Sci S A Lif Sci* 19(2): 146-151.
[13] Li, Y., & Ye, D. (2010). Cancer therapy by targeting hypoxia-inducible factor-1. *Current cancer drug targets,* 10(7), 782-796.
[14] Marshall, H. E., Hess, D. T., & Stamler, J. S. (2004). S-nitrosylation: Physiological regulation of NF-κB. *Proceedings of the National Academy of Sciences of the United States of America*, 101(24), 8841-8842. doi:10.1073/pnas.0403034101.
[15] Rahman, M., Senga, T., Ito, S., Hyodo, T., Hasegawa, H., Hamaguchi, M. (2010) S-Nitrosylation at Cysteine 498 of c-Src Tyrosine Kinase Regulates Nitric Oxide-mediated Cell Invasion. *Journal of Biological Chemistry* 285: 3806-3814.
[16] Sha, Y., Marshall, H. (2012). S-nitrosylation in the regulation of gene transcription. *Biochim Biophys Acta* 1820: 701-711.
[17] Schonhoff, C. M., Daou, M. C., Jones, S. N., Schiffer, C. A., Ross, A. H. (2002). Nitric Oxide-Mediated Inhibition of Hdm2-p53 Binding. *Biochemistry* 41(46): 13570-13574.
[18] Singh, S., Gupta, A. K. (2011). Nitric oxide: role in tumour biology and iNOS/NO-based anticancer therapies. *Cancer Chemoth Pharm* 67 (6): 1211-1244.

[19] Tang, C., Wei, W., Liu, L. (2012). Regulation of DNA repair by S-nitrosylation. *Biochim Biophys Acta* 1820: 730-735.

[20] Wang, Z. (2012). Protein S-nitrosylation and cancer. *Cancer Lett* 320: 123-129.

[21] Chan, C.-H., Jo, U., Kohrman, A., Rezaeian, A. H., Chou, P.-C., Logothetis, C., & Lin, H.-K. (2014). Posttranslational regulation of Akt in human cancer. *Cell & Bioscience*, 4(1), 59.doi:10.1186/2045-3701-4-59.

[22] Shi, Y, Paluch, B. E., Wang, X, Jiang, X. (2012). PTEN at a glance. *Journal of Cell Science* 125: 4687-4692.

[23] Calmels, S., Hainaut, P., Shshima, H. (1997). Nitric oxide induces conformational and functional modifications of wild-type tumor suppressor protein. *Cancer Res* 57: 3365-3369.

[24] Jaiswal, M., LaRusso, N. F., Nishioka, N., Nakabeppu, Y., Gores, G. J. (2001). HumanOgg1, a protein involved in the repair of 8-oxoguanine, is inhibited by nitric oxide. *Cancer Res.* 61: 6388-6393.

[25] Keyse, S. M. (2008). Dual-specificity MAP kinase phosphatases (MKPs) and cancer. *Cancer and Metastasis Reviews* 27: 253-261.

[26] Karp, G. (2013). Cancer. *Cell and Molecular Biology: Concepts and Experiments* 7:664-698.

[27] Kang-Decker, N., Cao, S., Chatterjee, S., Yao, J., Egan, L. J., Semela, D., Mukhopadhyay, D., Shah, V. (2007). Nitric oxide promotes endothelial cell survival signaling throughS-nitrosylation and activation of dynamin-2. *Journal of Cell Science* 120: 492-501.

In: Nitric Oxide
Editor: Dominic M. Evans
ISBN: 978-1-63482-961-8

Chapter 3

NITRIC OXIDE AND TRYPANOSOMATIDS: A TALE OF DR. JEKYLL AND MR. HYDE

***D. L. Fabrino*[1] *and L. L. Leon*[2,*]**
[1]Department of Chemistry,
Biotechnology and Bioprocess Engineering,
UFSJ/CAP, Brazil
[2]Laboratory of Biochemistry of Trypanosomatids,
Oswaldo Cruz Institute, FIOCRUZ, Brazil

ABSTRACT

Nitric Oxide (NO) is a highly reactive and unstable free radical that is produced by L-arginine oxidation to citrulline mediated by the enzyme Nitric Oxide Synthase (NOS). NO has been considered a key molecule on innate immune responses against different parasites, including *Trypanosoma cruzi* (*T. cruzi*), however a careful look at the literature about the relationship of NO and *T. cruzi* immune response shows many controversial data about it. *T. cruzi* is an obligate intracellular parasite with a complex life cycle involving invertebrates and vertebrates and is the causal agent of Chagas disease, considered one of the most important public health issues in Latin America, where several million people are infected. Different escape mechanisms have been developed by parasites thanks to their long co-evolution with host cells which exhibit many regulatory mechanisms manipulated by the parasites. Regarding NO, this

* Corresponding author: lleon@globo.com.

is not different. Therefore, it is important to understand these mechanisms in order to develop control strategies to different infectious pathologies. Two particular aspects have called our attention: first, many parasites have NOS and exhibit the ability to produce NO including trypanosomatids; and second, even if they do not manipulate the metabolic pathway of NO production in the host cell, they are able to resist to high levels of this molecule. Despite being a potent mediator molecule with multiple functional activities as extensively demonstrated in mammalian cells, the occurrence of NO in protozoan parasites and its physiological role has received much less attention. Also, the ability to resist to high levels of NO during the infectious process has not been much exploited. The use of chemical NO donors, such as sodium nitroprusside can be helpful to evaluate both the resistance of microorganisms to this molecule and the amount of NO needed to cause toxicity. This chapter aims to discuss the complexity of the NO metabolic pathway in order to better understand its role in the interaction between parasites and host cells.

INTRODUCTION

The presence of free radicals in biological systems was discovered around the 1960's. These molecules were immediately related to prejudicial processes in the organisms, such as intense cellular damage, mutation and cell ageing. In the 1980's the literature already exhibited papers demonstrating the beneficial roles of these molecules in cell signaling for instance (Droge, 2002).

Among these contradictory free radicals were the nitrogenous oxides with the Nitric Oxide (NO) standing out, once it was identified as a center piece of both immune responses (Drapier et al., 1988; Vincendeau & Daulouede, 1991; Gazzinelli et al., 1992; Oswald et al., 1994; Coleman, 2001; Sosroseno et al., 2004) and homeostatic reactions (Kroncke et al., 1997; Felley-Bosco et al., 2002; Dusse et al., 2003; Endres et al., 2004; Martinez & Andriantsitohaina, 2008; Abe et al., 2009).

Actually what is crucial for determination of the benefic or malefic role of NO in mammals is the balance among its concentration and the concentration of pro inflammatory cytokines (Brunet, 2001; Martinez & Andriantsitohaina, 2008), thus it is reasonable to think that the same is valid for different microorganisms which exhibit this molecule in its metabolism.

NO is a highly reactive and unstable free radical that is produced by L-arginine oxidation to citrulline mediated by the enzyme Nitric Oxide Synthase (NOS) (Rivero, 2006). NOS have been purified from a variety of mammalian tissues and has also been reported to be present in lower eukaryotic organisms (Chen & Rosazza, 1994; Nathan & Xie, 1994; Johansson & Carlberg, 1995; Hobbs & Ignarro, 1996; Basu et al., 1997; Guedes-Silva et al., 2006). It is a molecule widely distributed as it can be found in plants (Romero-Puertas et al., 2004), insects (Rivero, 2006), lower eukaryotes such as *T. cruzi* epimastigote forms (Paveto et al., 1995) and *Leishmania* sp. (Basu et al., 1997; Geigel & Leon, 2003; Guedes-Silva, et al., 2006), fungus (Kanadia et al., 1998) and *T. cruzi* trypomastigote and amastigote forms (Fabrino, 2009) or even in prokaryotes (Corker & Poole, 2003; Zumft, 2005).

As a result, the metabolic pathway of NO has been wearily studied and it is very well described, especially in mammals. Nonetheless, in this pathway there are many intersection points with other metabolic routes that create a complexity of situations that can favor, or not, the interaction between parasites and host cells, discussed in more detail in the chapter "L-arginine metabolism in pathogenic trypanosomatids *Trypanosoma cruzi* and *Leishmania* sp." in this book.

A careful look in the literature about the relationship of NO and *T. cruzi* immune response shows many controversial data about it. Moreover, there are data showing that parasites can even use this molecule in their metabolism, as a signal molecule; so it seems that this molecule lives the tale of Dr. Jekyll and Mr. Hyde not only in mammals, as is so well reported in the literature and pointed out in this text, but also in the trypanosomatids, especially in *T. cruzi.*

T. cruzi is an obligate intracellular parasite with a complex life cycle involving two different hosts, one invertebrate, the Reduviidae bugs generally called barber bug; and the vertebrate, which can be several different mammals, including humans. When the infection is established in humans it will cause the pathology named Chagas disease, considered one of the most important public health problems in Latin America, where several million people are infected (Barrett et al., 2003).

The parasite life cycle involves an extracellular, proliferative stage (epimastigote) that resides in the insect vector, and two forms that occur in the mammalian host, a non-proliferating, infective form (trypomastigotes) and an intracellular, proliferative form (amastigotes) (Irigoin et al., 2008).

Immune control of *T. cruzi* has been linked to a dominant Th_1 response, leading to macrophage activation and elimination of intracellular amastigotes through the induction of nitric oxide synthase (NOS II) and nitric oxide (NO) synthesis from L-arginine (Vespa et al., 1994; Silva et al., 1995; Rodrigues et al., 2000; Denkers & Butcher, 2005).

However, it is necessary to have an open mind about this issue as it is known that in different infection models the amount of NO produced by the host cells may vary according to the type of the invading parasite (Brunet, 2001). The adage first expressed by Paracelsus, founder of toxicology, 'The dose makes the poison' is logically valid for NO as discussed by Nagy 2007 (Nagy et al., 2007). This way, it is mandatory to keep in mind that every cell/tissue has its own threshold of resistance to NO (Kroncke et al., 1997) and we believe that this threshold may also vary according to the cell life stage, as will be discussed better ahead in this chapter. Despite the hazard offered by the NO molecule, well described in the literature (Oswald et al., 1994; James, 1995; Guzik et al., 2003; Peluffo et al., 2004; Genestra, 2007), different escape mechanisms were developed by parasites thanks to their long co-evolution with host cells (Rivero, 2006), and these cells exhibit many regulatory mechanisms to be manipulated by the parasites, which is no different regarding the NO pathway.

In vivo studies have demonstrated that the parasite is able to manipulate the NO pathway of the host cell and resist to the NO attack (Freire-de-Lima et al., 2000; Piacenza et al., 2001; Peluffo et al., 2004). Given the necessity to avoid the toxic effects of macrophage-generated NO, different hypotheses might account for the differential survival of the different forms of *T. cruzi* (Mauel, 1996; Denkers & Butcher, 2005; Irigoin et al., 2008).

For instance, it has been proposed that the parasite is able to induce the expression of arginase in the host cell, an enzyme which competes with nitric oxide synthase (NOS) by its substrate, the L-arginine. The induction of arginase would lead to lower levels of NO and, in parallel, higher production of polyamines, the final metabolite of arginase. Polyamines are used for the parasite growth and differentiation (Vincendeau et al., 2003; Peluffo et al., 2004).

Another possible mechanism involved in the decrease of NO levels released by the host cell, is the modulation of its Th_1/Th_2 cytokines profile, by the parasite. During the infection of macrophages with *T. cruzi*, the engulfment of apoptotic T cells leads to the production of TGFβ, a known down regulator of NO production (Freire-de-Lima et al., 2000). However, how the parasite survives to the NO hostile response is not well understood.

Therefore, it is important to understand the metabolic route of NO not only in the hosts but also in the parasites in order to develop control strategies to different infectious pathologies. Two particular aspects have called our attention on this topic: first, many parasites have NOS and exhibit the ability to produce NO, including trypanosomatids, and second, different parasites especially trypanosomatids are able to resist to high levels of this molecule.

In mammals, the NOS is recognized as responsible for NO production, and presents three isoforms (nNOS = neuronal NOS, eNOS = endothelial NOS and iNOS = inducible NOS). All the enzymes characterized so far are hemoprotein dimers comprised of subunits and require cofactors such as NADPH, tetrahydrobiopterin, FAD and FMN. The C-terminal domain is considered homologous among the isoforms, existing lesser homology in N-terminal domain. NOS is a modular enzyme in which each monomer consists of several discrete domains (Stamler et al., 1992; Stuehr, 1999; Witteveen et al., 1999; Alderton et al., 2001).

Moreover, NOS has been purified from a variety of mammalian tissues and has also been reported to be present in lower eukaryotic organisms (Chen & Rosazza, 1994; C. Nathan & Xie, 1994; Johansson & Carlberg, 1995; Hobbs & Ignarro, 1996; Basu et al., 1997; Guedes-Silva et al., 2006).

NO is an intriguing and potent mediator molecule with multiple functional activities as extensively demonstrated in mammalian cells (revision in (Martinez & Andriantsitohaina, 2008). The occurrence of this molecule in protozoan parasites, however, has received much less attention. To date, it has been described only in members of the genus *Leishmania* (Basu et al., 1997; Geigel & Leon, 2003; Guedes-Silva et al., 2006) and in epimastigote forms of *T. cruzi* (Paveto et al., 1995). The occurrence of NOS in epimastigotes of *T. cruzi*, was reported for the first time in 1995 by Paveto and collaborators. They showed not only the presence of a Ca^{2+}/dependent NOS but also evidenced a NO signaling pathway in this protozoan. It was also demonstrated the existence of a nitroprusside-activated guanylyl cyclase and N-methyl-D-aspartate (NMDA) receptors in epimastigote forms of this parasite (Paveto et al., 1995).

However, the ability of *T. cruzi* forms (trypomastigotes and amastigotes) present in mammalian host to synthesize NO remains to be defined in the literature. Our group has worked on this issue and as it was shown in Fabrino's PhD thesis this parasite is able to produce NO, interestingly the trypomastigotes are able to release significant higher levels (45%) of NO than amastigotes (Fabrino, 2009). As aforementioned, NO apparently affects a wide range of cellular functions in mammals such as cell-to-cell interactions, cellular proliferation, cytotoxicity and cytostasis, homeostatic and numerous other physiological mechanisms (Bredt & Snyder, 1994; Droge, 2002; Martinez & Andriantsitohaina, 2008). Moreover, several works indicated NO as an important antimicrobial molecule (Nathan, 1982; Clark & Rockett, 1996;

Kroncke et al., 1997; Bogdan et al., 2000). In invertebrates, however, little is known about the functional activity of this molecule.

It has been suggested that the stimulation of the soluble guanylate cyclase (GCs) activity by NO, would be responsible for the increase of the motility of the flagellum of *T. cruzi* epimastigotes (Pereira et al., 1997). Different purposes of this pathway in the parasite cell can be postulated. As proposed to the epimastigote forms, the greater activity of NOS and higher levels of released NO by trypomastigotes, might also be related to the high motility of these forms. Trypomastigotes are considered the main infective forms of *T. cruzi*, reaching the host cell through the blood stream and infecting them; consequently, it is reasonable to think that the higher level of NO produced by trypomastigotes is linked to the flagellum movements.

Once trypomastigotes are the parasite forms which get in contact with the harsh environment of the phagosome, it might be possible that a negative feedback can be driven by the NO produced by the parasite on the iNOS from the host cell. This mechanism has also been suggested in the literature when high levels of NO are produced by iNOS from host cell (Brunet, 2001) and the parasite might be using its produced NO to protect itself from the apoptosis process. The presence of a caspase-like activity was shown in *T. cruzi* epimastigotes suggesting a possible enzymatic step that could be down regulated by NO through S-nitrosylation (reviewed in Piacenza et al., 2001, 2002). The scheme bellow shows some of these functions.

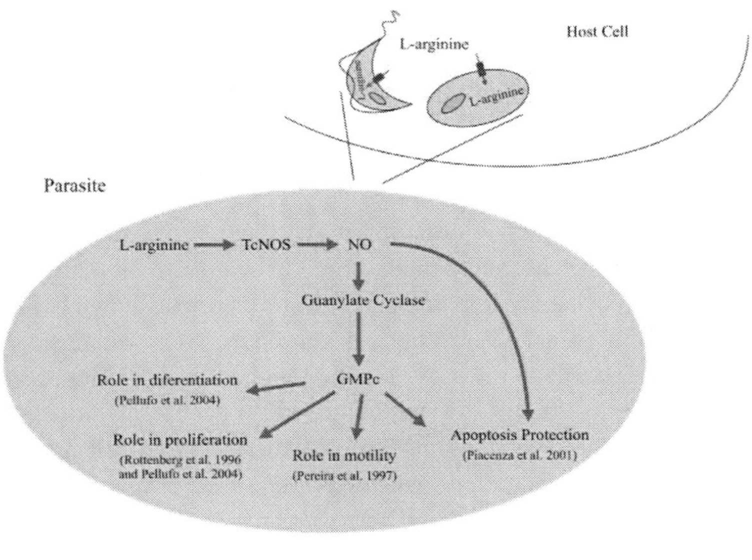

Figure 1. NO possible actions in *T. cruzi* homeostasis (Fabrino, 2009).

It is also worth to think about the environment of each stage of the parasite, for instance amastigotes have muscle cells as one of the favorite cell types for the maintenance of infection as well as macrophages; both differentiated cells, thus they have lower rates of arginine to offer to the proliferating cells (Peluffo et al., 2004). The available substrate is a rate limiting factor for the NOS (Alderton et al., 2001), if we consider that the macrophage spends a great amount of L-arginine in the first hours of infection in the process called oxidative burst (Bogdan, 1997) and that the trypomastigotes forms would also require the L-arginine in order to produce its own NO. As suggested by Peluffo and collaborators (2004) it is possible to hypothesize that the amastigotes not only are living in a poorer L-arginine environment but also driving the major part of the obtained L-arginine for multiplication.

As pointed out before, the L-arginine metabolism by NO pathway can also be protective of *T. cruzi* epimastigote forms against apoptotic death, triggered by human fetal serum. In an elegant study, Rafael Radi's group (2001) first induced *T. cruzi* epimastigote forms to massive apoptosis with human fetal serum (10%) in culture and observed that if the parasites were first incubate with L-arginine, the substrate for both NO and polyamines pathways, there was no DNA fragmentation; furthermore, a subsequent incubation with L-NAME, a NOS inhibitor, lead to intense DNA fragmentation suggesting the NO cytoprotector role in this parasite (Piacenza et al., 2001). The same research group in 2002 provided chemical donors of NO to *T. cruzi* cultures and were able to recover the parasites multiplication rates (Piacenza et al., 2002).

Moreover, the literature suggests that trypomastigotes are able to elicit asynchronous and repetitive intracellular Ca^{2+} pulses in the host cell, such changes regulate parasite invasion by promoting transient actin rearrangements and lysosome-plasma membrane fusion and subsequently the level of this ion declines at later times of infection (Ruiz et al., 1998; Burleigh & Woolsey, 2002). This calcium kinetics could be interfering in the trypomastigotes and amastigotes ability to produce NO once the literature (Paveto et al., 1995) and our unpublished data suggests *T. cruzi* may exhibit a constitutive Ca^{2+} dependent TcNOS.

Nonetheless, further studies are necessary to investigate if the NO produced by the parasite can modulate the production of Th_1/Th_2 cytokines during the infection. Although macrophages trigger their defense mechanism to neutralize the parasite, our results have shown that the *T cruzi* NO pathway is a consistent system, differentially expressed in the parasite forms, and like in *Leishmania,* might modulate host NO/NOS production, infection range or other intracellular signaling pathways (Pereira et al., 1997; Temporal et al., 2005; Genestra et al., 2006). However, despite not manipulating the metabolic pathway of NO production in the host cell, *T. cruzi* is able to resist to high levels of this molecule.

The use of chemical NO donors such as the sodium nitroprusside (Na_2 [Fe $(CN)_5NO].2H_2O$), (SNP) is a helpful tool to evaluate both the resistance of microorganisms to this molecule and the amount of NO needed to cause toxicity (Yamamoto & Bing, 2000; Rivero, 2006; Genestra et al., 2008). NO donors are pharmacologically active substances, with a functional active nitrate and nitroso groups within the molecule. These compounds are able to release NO both, *in vivo* and *in vitro*, (Bates et al., 1991). SNP is a complex of ferrous ion with five cyanide anions (CN^-) and a nitrosonium (NO^+). Interaction of SNP with a reducing agent such as a thiol leads to the formation of NO (Bates et al., 1991; Rao & Cederbaum, 1995; Ramakrishna Rao & Cederbaum, 1996; Bitar et al., 2005).

Using the SNP against *Leishmania amazonensis*, *in vitro*, our group has demonstrated that SNP induced loss of viability in both promastigotes and axenic amastigotes, being the latter parasite forms statistically more resistant to the effects of NO than the promastigotes (Genestra et al., 2008).

For the first time the use of the SNP and the effects of different doses of NO targeted at the trypomastigote and amastigote forms of *T. cruzi*, *in vitro* was observed by our group. The results demonstrated that NO is not such a straight trypanocidal molecule, at least not for the amastigote forms, a differential resistance of this parasite forms to NO with higher (53%) was seen as compared to trypomastigotes, suggesting the possible participation of NO in the differentiation process of this protozoan (Fabrino, 2009).

The observation that *T. cruzi* amastigotes are quite more resistant to NO than trypomastigotes is in agreement with a previous work with *L. amazonensis,* in which amastigotes were pointed as the more resistant life stage of this intracellular pathogen.

When treated with the same doses, of the same NO donor used for *T. cruzi* in the above cited work, promastigotes of *L. amazonenzis* showed dramatic loss of viability (99% at 8.0μM of SNP) whereas on amastigotes forms the loss of viability was about 50%. Promastigotes went through dramatic morphological changes, such as loss of the apparent flagellum and changing in their shape (from elongated to rounded) (Genestra et al., 2008).

The same was observed under light microscopy for *T. cruzi* trypomastigotes forms, which seem to undergo an amastigote-like form after being in contact with NO. These changes in morphology occur in a dose dependent way, once at higher doses it was found a higher number of rounded shape forms lacking an apparent flagellum. These observations added to the ability of amastigotes to survive at high doses of NO, allows speculation on a possible role for NO in the differentiation process of this parasite.

This way, notwithstanding the fact that NO is considered a pivotal killing molecule against parasites (James, 1995; Clark & Rockett, 1996; Bogdan et al., 2000), including *T. cruzi* (Gazzinelli et al., 1992; Vespa et al., 1994; Petray et al., 1995; Silva et al., 1995; Aliberti et al., 1996; Holscher et al., 1998; Rodrigues et al., 2000), there are also data in the literature proposing its homeostatic, and even a protective, role in trypanosomatids such as in *Leishmania* sp., in which the NO could play a role in its infectivity (Temporal et al., 2005) and *T. cruzi,* as discussed above.

Lastly, it is worth to note that the order Kinetoplastidae, to which *T. cruzi* belongs, contains other protozoans presenting very similar genomic and structural organization (Stuart et al., 2008). Taking into consideration that about 6000 to 8000 genes of this protozoan are orthologues and that the Kinetoplastidae are among the most ancient eukaryotes, it is natural that they present several similar biochemical pathways and it is probable that they even share some pathways with their hosts (Donelson et al., 1999; Rivero, 2006; Stuart et al., 2008).

Conclusion

The presence of the same enzyme, NOS; the same ability to resist to its product NO and the possibility of using this molecule for physiological functions in two protozoans which share many metabolic routes besides their biological and morphological characteristics, make us re-think the real role of NO in these parasites.

Is this molecule the doctor or the monster?

REFERENCES

Abe Y., Hines I., Zibari G. & Grisham M. B. (2009), Hepatocellular protection by nitric oxide or nitrite in ischemia and reperfusion injury. *Arch Biochem Biophys,* 484, 232-237.

Alderton W. K., Cooper C. E., Knowles R. G. (2001). Nitric oxide synthases: structure, function and inhibition. *Biochem J,* 357, 593-615.

Aliberti J. C., Cardoso M. A., Martins G. A., Gazzinelli R. T., Vieira L. Q., Silva J. S. (1996). Interleukin-12 mediates resistance to *Trypanosoma cruzi* in mice and is produced by murine macrophages in response to live trypomastigotes. *Infect Immun,* 64, 1961-1967.

Barrett M. P., Burchmore R. J., Stich A., Lazzari J. O., Frasch A. C., Cazzulo J. J., Krishna S. (2003) The trypanosomiases. *Lancet,* 362, 1469-1480.

Basu N. K., Kole L., Ghosh A., Das P. K. (1997), Isolation of a nitric oxide synthase from the protozoan parasite, *Leishmania donovani. FEMS Microbiol Lett,* 156, 43-47.

Bates J. N., Baker M. T., Guerra R. Jr., Harrison D. G. (1991), Nitric oxide generation from nitroprusside by vascular tissue. Evidence that reduction of the nitroprusside anion and cyanide loss are required, *Biochem Pharmacol,* 42 Suppl, S157-165.

Bitar M. S., Wahid S., Mustafa S., Al-Saleh E., Dhaunsi G. S., Al-Mulla F. (2005). Nitric oxide dynamics and endothelial dysfunction in type II model of genetic diabetes. *Eur J Pharmacol,* 511, 53-64.

Bogdan C. (1997). Of microbes, macrophages and nitric oxide, *Behring Inst Mitt*, 99, 58-72.

Bogdan C., Rollinghoff M., Diefenbach A. (2000.) Reactive oxygen and reactive nitrogen intermediates in innate and specific immunity. *Curr Opin Immunol,* 12, 64-76.

Bredt D., Snyder S. H. (1994). Nitric oxide: a physiologic messenger molecule. *Annu Rev Biochem,* 63, 175-195.

Brunet L. R. (2001). Nitric oxide in parasitic infections. *Int Immunopharmacol,* 1, 1457-1467.

Burleigh B. A., Woolsey A. M. (2002). Cell signalling and *Trypanosoma cruzi* invasion. *Cell Microbiol,* 4, 701-711.

Chen Y., Rosazza J. P. (1994). A bacterial nitric oxide synthase from a *Nocardia* species. *Biochem Biophys Res Commun,* 203, 1251-1258.

Clark I. A., Rockett K. A. (1996). Nitric oxide and parasitic disease. *Adv Parasitol,* 37, 1-56.

Coleman J. W. (2001). Nitric oxide in immunity and inflammation. *Int Immunopharmacol,* 1, 1397-1406.

Corker H., Poole R. K. (2003). Nitric oxide formation by *Escherichia coli.* Dependence on nitrite reductase, the NO-sensing regulator Fnr, and flavohemoglobin Hmp. *J Biol Chem,* 278, 31584-31592.

Denkers E. Y., Butcher B. A. (2005). Sabotage and exploitation in macrophages parasitized by intracellular protozoans. *Trends Parasitol,* 21, 35-41.

Donelson J. E., Gardner M. J., El-Sayed N. M. (1999). More surprises from Kinetoplastida. *Proc Natl Acad Sci USA,* 96, 2579-2581.

Drapier J. C., Wietzerbin J., Hibbs J. B. Jr. (1988). Interferon-gamma and tumor necrosis factor induce the L-arginine-dependent cytotoxic effector mechanism in murine macrophages. *Eur J Immunol,* 18, 1587-1592.

Droge W. (2002). Free radicals in the physiological control of cell function. *Physiol Rev,* 82, 47-95.

Dusse L. M. S., Vieira L. M., Carvalho M. G. (2003). Revisão sobre óxido nítrico. *Jornal Brasileiro de Patologia e Medicina Laboratorial,* 39, 8.

Endres M., Laufs U., Liao J. K., Moskowitz M. A. (2004). Targeting eNOS for stroke protection. *Trends Neurosci,* 27, 283-289.

Fabrino D. L. (2009), Nitric oxide and its relationship with different microorganisms *in vitro* and during experimental acute infection with *Trypanosoma cruzi*., Oswaldo Cruz Institute, Rio de Janeiro.

Felley-Bosco E., Bender F., Quest, A. F. (2002), Caveolin-1-mediated post-transcriptional regulation of inducible nitric oxide synthase in human colon carcinoma cells, *Biol Res,* 35, 169-176.

Freire-de-Lima C. G., Nascimento D. O, Soares M. B., Bozza P. T., Castro-Faria-Neto H. C., de Mello F. G., Dos Reis G. A., Lopes M. F. (2000). Uptake of apoptotic cells drives the growth of a pathogenic trypanosome in macrophages. *Nature,* 403, 199-203.

Gazzinelli R. T., Oswald I. P., Hieny S., James S. L., Sher A. (1992). The microbicidal activity of interferon-gamma-treated macrophages against *Trypanosoma cruzi* involves an L-arginine-dependent, nitrogen oxide-mediated mechanism inhibitable by interleukin-10 and transforming growth factor-beta. *Eur J Immunol,* 22, 2501-2506.

Geigel L. F., Leon L. L. (2003). Cyclic 3'-5' guanosine monophosphate-dependent activity in *Leishmania amazonensis. Mem Inst Oswaldo Cruz,* 98, 499-500.

Genestra M. S., Guedes-Silva D., Souza W. J., Cysne-Finkelstein L., Soares-Bezerra R. J., Monteiro F. P., Leon L. L. (2006). Nitric oxide synthase

(NOS) characterization in *Leishmania amazonensis* axenic amastigotes, *Arch Med Res,* 37, 328-333.

Genestra M. S., Souza W. J., Guedes-Silva D., Machado G. M., Cysne-Finkelstein L., Bezerra R. J., Monteiro F., Leon L. L. (2006), Nitric oxide biosynthesis by *Leishmania amazonensis* promastigotes containing a high percentage of metacyclic forms. *Arch Microbiol.* 185, 348-354.

Genestra M. S. (2007), Oxyl radicals, redox-sensitive signalling cascades and antioxidants. *Cell Signal,* 19, 1807-1819.

Genestra M. S., Soares-Bezerra R. J., Gomes-Silva L., Fabrino D. L., Bellato-Santos T., Castro-Pinto D. B., Canto-Cavalheiro M. M., Leon L. L. (2008). *In vitro* sodium nitroprusside-mediated toxicity towards *Leishmania amazonensis* promastigotes and axenic amastigotes, *Cell Biochem Funct,* 26, 709-717.

Guzik T. J., Korbut R., Adamek-Guzik T. (2003), Nitric oxide and superoxide in inflammation and immune regulation. *J Physiol Pharmacol,* 54, 469-487.

Hobbs A. J., Ignarro L. J. (1996), Nitric oxide-cyclic GMP signal transduction system. *Methods Enzymol,* 269, 134-148.

Holscher C., Kohler G., Muller U., Mossmann H., Schaub G. A., Brombacher F. (1998). Defective nitric oxide effector functions lead to extreme susceptibility of *Trypanosoma cruzi*-infected mice deficient in gamma interferon receptor or inducible nitric oxide synthase. *Infect Immun,* 66, 1208-1215.

Irigoin F., Cibils L., Comini M. A., Wilkinson S. R., Flohe L., Radi R. (2008). Insights into the redox biology of *Trypanosoma cruzi*: Trypanothione metabolism and oxidant detoxification. *Free Radic Biol Med,* 45, 733-742.

James S. L. (1995), Role of nitric oxide in parasitic infections. *Microbiol Rev,* 59, 533-547.

Johansson K. U., Carlberg M. (1995). NO-synthase: what can research on invertebrates add to what is already known? *Adv Neuroimmunol,* 5, 431-442.

Kanadia R. N., Kuo W. N., McNabb M. Botchway A. (1998), Constitutive nitric oxide synthase in Saccharomyces cerevisiae. *Biochem Mol Biol Int,* 45, 1081-1087.

Kroncke K. D., Fehsel K., Kolb-Bachofen V. (1997). Nitric oxide: cytotoxicity versus cytoprotection: How, why, when, and where? *Nitric Oxide,* 1, 107-120.

Martinez M. C., Andriantsitohaina R. (2008). Reactive Nitrogen Species: Molecular Mechanisms and Potential Significance in Health and Disease. *Antioxid Redox Signal.*

Mauel J. (1996). Intracellular survival of protozoan parasites with special reference to *Leishmania* sp., *Toxoplasma gondii* and *Trypanosoma cruzi. Adv Parasitol,* 38, 1-51.

Nagy G., Clark J. M., Buzas E. I., Gorman C. L., Cope A. P. (2007). Nitric oxide, chronic inflammation and autoimmunity. *Immunol Lett,* 111, 1-5.

Nathan C., Xie Q. W. (1994). Nitric oxide synthases: roles, tolls, and controls. *Cell,* 78, 915-918.

Nathan C. F. (1982). Secretion of oxygen intermediates: role in effector functions of activated macrophages, *Fed Proc,* 41, 2206-2211.

Oswald I. P., Wynn T. A., Sher A., James S. L. (1994). NO as an effector molecule of parasite killing: modulation of its synthesis by cytokines. *Comp Biochem Physiol Pharmacol Toxicol Endocrinol,* 108, 11-18.

Paveto C., Pereira C., Espinosa J., Montagna A. E., Farber M., Esteva M. Flawiá M. M., Torres H. N. (1995), The nitric oxide transduction pathway in *Trypanosoma cruzi. J Biol Chem,* 270, 16576-16579.

Peluffo G., Piacenza L., Irigoin F., Alvarez M. N., Radi R. (2004), L-arginine metabolism during interaction of *Trypanosoma cruzi* with host cells. *Trends Parasitol,* 20, 363-369.

Pereira C., Paveto C., Espinosa J., Alonso G., Flawia M. M., Torres H. N. (1997), Control of *Trypanosoma cruzi* epimastigote motility through the nitric oxide pathway. *J Eukaryot Microbiol,* 44, 155-156.

Petray P., Casta¤os-Velez E., Grisnstein S., Orn A., Rottenberg M. E. (1995), Role of nitric oxide in resistance and histophatology during experimental infection with *Trypanosoma cruzi, Immunol Letters,* 47, 121-126.

Piacenza L., Peluffo G., Radi R. (2001). L-arginine-dependent suppression of apoptosis in *Trypanosoma cruzi*: Contribution of the nitric oxide and polyamine pathways. *Proc Natl Acad Sci USA,* 98, 7301-7306.

Piacenza L., Peluffo G., Radi R. (2002). L-arginine metabolism in *Trypanosoma cruzi* in the regulation of programmed cell death. *Methods Enzymol,* 359, 286-302.

Ramakrishna Rao D. N., Cederbaum A. I. (1996), Generation of reactive oxygen species by the redox cycling of nitroprusside. *Biochim Biophys Acta,* 1289, 195-202.

Rao D. N., Cederbaum A. I. (1995), Production of nitric oxide and other iron-containing metabolites during the reductive metabolism of nitroprusside by microsomes and by thiols. *Arch Biochem Biophys,* 321, 363-371.

Rivero A. (2006). Nitric oxide: an antiparasitic molecule of invertebrates. *Trends Parasitol,* 22, 219-225.

Rodrigues M. M., Ribeirao M., Boscardin S. B. (2000). CD_4 Th_1 but not Th_2 clones efficiently activate macrophages to eliminate *Trypanosoma cruzi* through a nitric oxide dependent mechanism. *Immunol Letters,* 73, 43-50.

Romero-Puertas M. C., Perazzolli M., Zago E. D., Delledonne M. (2004). Nitric oxide signalling functions in plant-pathogen interactions. *Cell Microbiol,* 6, 795-803.

Ruiz R. C., Favoreto S. Jr., Dorta M. L., Oshiro M. E., Ferreira A. T., Manque P. M., Yoshida N. (1998). Infectivity of *Trypanosoma cruzi* strains is associated with differential expression of surface glycoproteins with differential Ca^{2+} signalling activity. *Biochem J,* 330 (Pt 1), 505-511.

Silva J. S., Vespa G. N. R., Cardoso M. A. G., Aliberti J. C. S., Cunha F. Q. (1995). Tumor necrosis factor alpha mediates resistance to *Trypanosoma cruzi* infection in mice by inducing nitric oxide production in infected gamma interferon - activated macrophages. *Infect Immun,* 63, 4862-4867.

Sosroseno W., Herminajeng E., Bird P. S., Seymour G. J. (2004), L-arginine-dependent nitric oxide production of a murine macrophage-like RAW 264.7 cell line stimulated with *Porphyromonas gingivalis* lipopolysaccharide. *Oral Microbiol Immunol,* 19, 65-70.

Stamler J. S., Singel D. J., Loscalzo J. (1992), Biochemistry of nitric oxide and its redox-activated forms. *Science,* 258, 1898-1902.

Stuart K., Brun R., Croft S., Fairlamb A., Gurtler R. E., McKerrow J., Reed S., Tarleton R. (2008). Kinetoplastids: related protozoan pathogens, different diseases. *J Clin Invest,* 118, 1301-1310.

Stuehr D. J. (1999), Mammalian nitric oxide synthases, *Biochim Biophys Acta,* 1411, 217-230.

Temporal R. M., Cysne-Finkelstein L., Echevarria A., Silva-Gonçalves A. J., Leon L. L., Genestra M. S. (2005), Amidine derivatives and *Leishmania amazonensis*: an evaluation of the effect of nitric oxide (NO) production on the parasite-macrophage interaction. *J Enzyme Inhib Med Chem,* 20, 13-18.

Vespa G. N. R., Cunha F. Q., Silva J. S. (1994), Nitric oxide is involved in control of *Trypanosoma cruzi:* Induced parasitemia and directly kills the parasite *in vitro*. *Infect Immun,* 62, 5177-5182.

Vincendeau P., Daulouede S. (1991). Macrophage cytostatic effect on *Trypanosoma musculi* involves an L-arginine-dependent mechanism. *J Immunol,* 146, 4338-4343.

Vincendeau P., Gobert A. P., Daulouede S., Moynet D., Mossalayi M. D. (2003), Arginases in parasitic diseases, *Trends Parasitol,* 19, 9-12.

Witteveen C. F., Giovanelli J., Kaufman S. (1999). Reactivity of tetrahydrobiopterin bound to nitric-oxide synthase. *J Biol Chem,* 274, 29755-29762.

Yamamoto T., Bing R. J. (2000), Nitric oxide donors. *Proc Soc Exp Biol Med,* 225, 200-206.

Zumft W. G. (2005), Nitric oxide reductases of prokaryotes with emphasis on the respiratory, heme-copper oxidase type. *J Inorg Biochem,* 99, 194-215.

In: Nitric Oxide
Editor: Dominic M. Evans

ISBN: 978-1-63482-961-8

Chapter 4

NANOMATERIALS FOR *IN VIVO* NO SENSING

João M. M. Leitão[1], Eliana F. C. Simões[2] and Joaquim C. G. Esteves da Silva[2*]

[1]Centro de Investigação em Química (CIQ), Faculty of Pharmacy, University of Coimbra, Polo das Ciências da Saúde, Coimbra, Portugal

[2]Centro de Investigação em Química (CIQ), Department of Chemistry and Biochemistry, Faculty of Sciences, University of Porto, Porto, Portugal

ABSTRACT

The *in vivo* NO sensing is of outstanding relevance for the comprehension of the diverse physiological and pathological process in which this species is involved. Nevertheless, *in vivo* NO sensing is a complex challenge due to its own physicochemical properties, antioxidant mechanisms and possible interferences. Some electrochemical and optical sensors have already been developed for *in vivo* NO sensing. Most of the electrochemical sensors are amperometric and most of the optical sensors are based on fluorescence. New materials are being developed for new NO electrochemical or optical sensors. Due to their intrinsic properties, the nanomaterials are one of the most promising emergent materials. This

* Corresponding Author address: Department of Chemistry and Biochemistry, Faculty of Sciences, University of Porto, Rua do Campo Alegre 687, Porto 4169-007, Portugal; Email:jcsilva@fc.up.pt.

chapter review the most promising nanomaterials for the development of electrochemically or optical sensors for *in vivo* NO sensing.

Keywords: NO, *in vivo,* sensing, nanomaterials, optical, electrochemical

1. INTRODUCTION

In vivo sensing [1, 2] and *in vivo* imaging [3, 4] are emerging fields with outstanding relevance that attract the interest of investigators of different scientific fields. Their importance has to do with the *in situ* diagnostic, therapeutic evaluation or study of diverse biological processes in health and disease. An *in vivo* sensor is constituted by a recognition selective phase connected to a transduction element. When the recognition element phase is a biologically component the sensor is designated as a biosensor. Several sensors or biosensors were designed for the *in vitro* sensing. Beside all the characteristics of an *in vitro* sensor, as the sensitivity and selectivity, an *in vivo* sensor must be nontoxic, biocompatible and stable over the time in the biological environment with special and/or temporal resolution. The great progress in the development of *in vivo* sensing is the translation from the *in vitro* to the *in vivo* assays. Indeed some biological processes could not be reproduced or measured *in vitro*.

Nitric oxide (NO) is one of the main reactive oxygen and nitrogen species (ROS/RNS) and one of the more relevant biological molecules in order to understand several physiological or pathological processes. Indeed NO is involved in a variety of physiological and pathological process having a beneficial and injurious role in the human being depending of is concentration level [5, 6]. The *in vivo* sensing of NO allows understanding the physiological or pathological processes in which is involved. Even so the *in vivo* NO sensing is not straightforward due to is short half live, rapid diffusion, usually low concentrations and high reactivity [6-8]. Several analytical methodologies have been developed for their detection and/or quantification. The principal analytical methodologies proposed for their quantification are based on electron paramagnetic resonance (EPR), UV-Visible, electrochemical, quimioluminescence and fluorescence [7, 8]. Most of the developed NO sensors are optical or electrochemical and are essentially based in amperometric or fluorescence analytical techniques, respectively [9, 10].

The definition of nanomaterial focuses mainly on two properties, the size (any material within 1-100 nm) and the size-dependent properties of a given material [11, 12]. In the group of nanomaterials are included the different nanoallotropic forms of carbon as the fullerene forms [fullerene (FL), carbon nanotubes (CNT), graphene (GP), nano diamonds (ND) and carbon dots (CD), crystalline semiconductors nanoparticles (quantum dots, QD), metals, dioxide titanium, polymer and silica nanoparticles. Due to their intrinsic properties the nanomaterials are one of the most promising emergent materials for the development of *in vivo* electrochemical or optical sensors [1, 9, 13, 14]. In the development of electrochemical amperometric sensors they can be used as the electrode itself, as the nano matrix of an active recognition compound or they could facilitate the electron transference. In the development of optical fluorescence sensors they could be quite useful due to the high quantum yields, high photo stability, high Stokes shift, excitation and emission wavelengths in the visible range, narrow emission bands and tunable photo luminescent properties.

In this chapter a revision of the main nanomaterials used in the development of NO electrochemical and optical sensors is presented. It will focus on already developed *in vivo* NO sensors. This chapter is structured taking into consideration the class of nanomaterial in the basis of the literature described *in vivo* NO sensors.

2. Carbon Nanotubes

Carbon nanotubes (CNT) is one of the nanoallotropic forms of carbon with a nano cylindrical structure arranged in condensed aromatic rings of graphene that could be found as single wall (SWCNT) or multi wall (MWCNT), if they are formed respectively by one or more than one graphene sheet [15]. Even if the toxicity of CNT could be controversial due to their own physicochemical properties namely, their high length-to-diameter with a high large surface areas, high mechanical strength, chemical and thermal stability and high optical absorbance in the near infrared (NIR), the CNT is one of the most used nanomaterial in the electrochemical *in vivo* sensing but also in the *in vivo* imaging [16]. Based on MWCNT four electrochemical sensors [17-20] and based on SWCNT one optical sensor [21] were developed as *in vivo* NO sensors. In Table 1 are presented the principal analytical performance characteristics of the already developed *in vivo* NO sensors based CNT electrochemical and optical sensors.

Table 1. Principal analytical performance characteristics of the developed *in vivo* NO electrochemical and optical CNT based sensors*

Ref.	LOD (μM)	LR (μM)	Sensitivity	Selectivity	Stability	Resolution (Spatial / Temporal)
Electrochemical						
[17]	0.025	0.25 - 1	1.72 nA/μM	Ascorbate; NO_2^-; Dopamine; DOPAC; H_2O_2; O_2	Stable	Limited
[18]	0.0076	0.019 - 54	1.73 μA/μM	Glucose; L-arginine; NO_2^-; Dopamine; L-cysteine; Uric acid; Ascorbic acid	Stable	---
[19]	0.0043	1 - 40	1.10 μA/μM	Glucose; L-arginine; Ascorbic acid; H_2O_2; $O_2^{\cdot-}$	Stable	---
[20]	0.028	0.22 - 120	0.25 μA/μM	L-arginine; NO_2^-; NO_3^-; Ascorbic acid; Dopamine; Several ions	Stable	---
Optical						
[21]	0.070	Exponential decay	---	$ONOO^-$; NO_2^-; NO_3^-; HNO; ClO^-; $HO^\cdot$; H_2O_2	Stable	Spatial / Temporal

* Ref. – References; LOD – Limit of detection; LR – Linear range; DOPAC – 3,4-dihydroxyphe-nylacetic acid.

In two of the developed *in vivo* NO electrochemical sensors non functionalized MWCNT were immobilized in chitosan (CS) [17, 18]. In the first, hemin and MWCNT were linked to the CS by a cross-linking procedure using N-hidroxysuccinimide (NHS) / 1-[3-(Dimethylamino)propyl]-3-ethyl-carbodiimide methiodide (EDC methiodide). A posterior electrodeposition of the obtained CS film was done on the surface of a carbon fiber microelectrode. Based on the electrocatalytic reduction of NO by the hemin, locally applied exogenous NO was detected by this sensor in a rat brain by squarewave voltammetry (SWV) at a reduction potential of -0.850 V (vs. Ag/AgCl reference electrode) in the presence of oxygen [17]. In the other, MWCNT and Au nanoparticles (AuNPs) were dispersed in CS. The obtained suspension was then deposited on a modified glassy carbon electrode (GCE) and covered with a nafion solution. The NO release in mouse liver, spleen and kidneys was evaluated by amperometry at an oxidation potential +0.85 V (vs. saturated calomel reference electrode, SCE) [18].

In the other two electrochemical *in vivo* NO sensors MWCNT were previous functionalized [19, 20]. The first one was functionalized with poly-5,2':5',2''-terthiophene-3'-carboxylic acid (PTTCA) [19] and the other with azocarmine B (ACB) [20]. The functionalized PTTCA-MWCNT were placed in a GCE modified with an AuNPs film. After electrode modification, microperoxidase, catalase and superoxide dismutase enzymes were covalently immobilized in the PTTCA-MWCNT-AuNPs film by a NHS / 1-Ethyl-3-(3-dimethylaminopropyl)carbodiimide (EDC) cross-linking procedure. The microperoxidase enzyme was used as catalyst of the reduction of NO and the catalase and superoxide dismutase enzymes were used in order to eliminate the interferences respectively of the H_2O_2 and $O_2^{\cdot -}$. The released NO was evaluated in rat liver samples, in stomach and intestinal cancer cells by amperometry at a reduction potential of -0.8 V (vs. Ag/AgCl) [19]. The functionalized ACB-MWCNT was deposited in a GCE and the electro-polymerization of ACB was done by cyclic voltametry (CV). The obtained film of polyaxocarmine B-ACB-MWCNT was then covered with nafion. The determination of NO was done in rat liver cells amperometrically at an oxidation potential of + 0.75 V(vs. SCE) [20].

The only developed *in vivo* NO optical sensor based on carbon nanomaterials is a NIR fluorescent sensor able to overcome the auto fluorescence, the possible scattering of living tissues and capable of tissue penetration. The NO evaluation is based in the quenching of fluorescence of the NIR spectra of SWCNT covered with 3,4-diaminophenyl functionalized dextran (DAP-dex) at an excitation wavelength of 785 nm [21]. The NO determination was performed in stimulated macrophage cells and in the abdomen of a mouse after the injection of a NO standard solution. In Figure 1 is presented one example of the *in vivo* NO electrochemical sensing (a) and of the *in vivo* NO optical sensing (b) based in CNT previous described.

Four other NO sensors based in CNT were developed with potential for *in vivo* NO sensing [22-25]. The electrochemical sensors are based on imprinted functional polymer-MWCNT-CS [22], in tetrasulfonated nickel phthalo-cyanine functionalized-SWCNT [23] and the in hematite (α-Fe_2O_3)-CNT-CS [24]. A NIR fluorescence sensor is based in the functionalization of SWCNT with a specific DNA sequence of d(AT)15 oligonucleotides selective to NO [25].

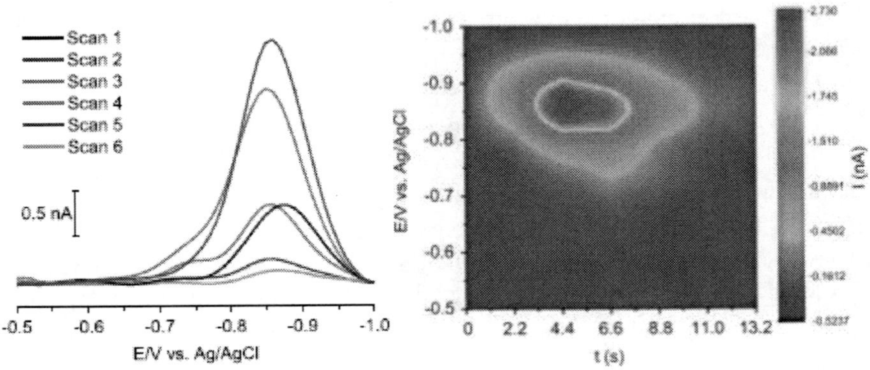

(a)

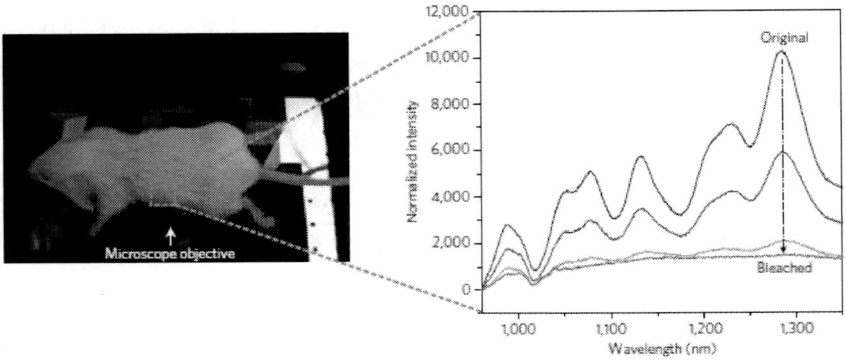

(b)

Figure 1. *In vivo* NO electrochemical sensing (a) by MWCNT-Hemin-CS electrodes in a rat brain (reproduced from reference [17] with permission from Elsevier) and *in vivo* NO optical sensing (b) by DAP-Dex-SWCNT in the abdomen of a mouse (reproduced from reference [21] with permission from the Nature Publishing Group).

3. Gold Nanoparticles

The AuNPs, also known as colloidal gold, are generally spherical nanoparticles usually synthesized in solution. Beside the size, the AuNPs also could differ in shape and can be found as nanospheres, nanorods, nanocages, nanocubes, nanopolygons or even nanostars. They have size dependent optical properties, large surface to volume ratio, biocompatibility and low toxicity, which make them useful materials for *in vitro* or *in vivo* sensing and for imaging [26, 27]. Beside the already indicated based CNT sensors, *in vivo* NO detection is based in AuNPs. Also, as for the CNT *in vivo* NO sensors, most of the developed AuNPs sensors are electrochemical. Based on AuNPs, and beside the previous indicated electrochemical MWCNT-AuNPs based electrodes [18, 19], four other electrochemical sensors [28-31] and three optical sensors [32-34] were proposed. In Table 2 are presented the principal analytical performance characteristics of the already developed *in vivo* NO based AuNPs electrochemical and optical sensors.

Table 2. Principal analytical performance characteristics of the developed *in vivo* NO electrochemical and optical AuNPs based sensors*

Ref.	LOD (μM)	LR (μM)	Sensitivity	Selectivity	Stability	Resolution (Spatial / Temporal)
Electrochemical						
[28]	0.1	0.4 - 673.9	3.653 mA. cm^{-2}/mM	NO_2^-; Dopamine; Uric acid; Ascorbic acid	---	---
[29]	0.0072	0.036 – 43.2	0.04 μA/μM	Uric acid; several ions	Stable	---
[30]	0.0036	0.0072 - 18	0.56 nA/nM	NO_2^-; Uric acid; Dopamine; Ascorbic acid; Glucose; Cholesterol; L-valine; L-arginine	Stable	---
[31]	0.05	0.1 - 40	---	NO_2^-; NO_3^-; Ascorbic acid; Uric acid; Neurotransmitters	Stable	---
Optical						
[32]	R: 8	R: < 700	R: 0.56 $(\Delta R_0/R)$/ mM	R: NO_2^-; NO_3^-; Cl^-; O_2; H_2O_2; $O_2^{\cdot-}$; $ONOO^-$	R:Stable	R:---
	τ: 30	τ: < 500	τ: 3.9 [Δτ (ns)]/mM	τ: ---	τ:Stable	τ:---
[33]	20	< 1000	0.23 $(\Delta R_0/R)$/ mM	H_2O_2; $O_2^{\cdot-}$; $ONOO^-$	Stable	---
[34]	20	< 1000	0.19 $(\Delta I_0/I)$/ mM	NO_2^-; NO_3^-; O_2; N_2; Ascorbic acid; H_2O_2; $O_2^{\cdot-}$; $ONOO^-$	Stable	---

* Ref. – References; LOD – Limit of detection; LR – Linear range; R – Ratiometric sensor; R_0 – Ratio without NO; R – Ratio at each NO concentration;τ - Lifetime sensor.

All of the developed *in vivo* NO electrochemical sensors based in AuNPs are amperometric [28-30]. In the first, Au and Pt nanoparticles were placed in a graphene paper electrode. The NO release in living cells were evaluated at an oxidation potential of + 0.75 V(vs. Ag/AgCl) [28]. In another, AuNPs and CS were deposited electrochemically at the surface of a GCE. The release of NO was then evaluated an oxidation potential of +0.85 V (vs. SCE) in rat kidney [29]. AuNPs were deposited in quartz fiber electrode, niacinamide (NA) was electropolymerized above the AuNPs film and posterior coated with nafion. The NO release from a mice liver were done at an oxidation potential of +0.65 V (vs. Ag/AgCl) [30]. In the last electrochemical sensor AuNPs were deposited at a surface of a Pt electrode and then coated with nafion. The NO release from smooth muscle cells was evaluated at an oxidation potential of + 0.74 V (vs. SCE) by differential pulse amperometry [31].

All of the developed *in vivo* NO optical sensors based in AuNPs are based in the fluorescence quenching by NO of a fluorophore previous linked to the heme protein cytochrome c (Cyt-c) highly selective to NO [32-34]. Based in these labeled Cyt-c optical fiber sensors were developed. The AuNPs were used for the attachment of the fluorophore labeled Cyt-c to the optical fibers. By the same authors the fluorophores used were the 4-carboxy-2',7'-difluorofluorescein, succinimidyl ester (Oregon Green 488 carboxylic acid, succinimidyl ester) [32, 33] and 4-carboxy-3,5,6 trifluoro-2',7'-difluorofluorescein, succinimidyl ester (Oregon Green 514 carboxylic acid, succinimidyl ester) [34]. Of the developed optical fiber sensors two are fluorescence ratiometric sensors [32, 33], one is a fluorescence lifetime sensor [32] and, the last one, a single fluorescence intensity determination at maximum fluorescence intensity [34]. For the developed fluorescence ratiometric sensors the detection of NO was done at an excitation wavelength of 488 nm and emission wavelengths of 540 and 545 nm. As fluorescence intensity reference, in the development of these fluorescence ratiometric sensors, it was used a fluorescent carboxylate-modified polystyrene nanospheres with a similar excitation wavelength of the fluorophore used (488 nm) and emission wavelength of 685 nm. For the developed fluorescence lifetime sensor, at an excitation of 488 nm, the lifetime was evaluated at an emission wavelength of 580 nm. The single fluorescence intensity evaluation was done at an emission wavelength of 640 nm and excitation of 530 nm. The NO extracellular release in macrophage cells was performed by the developed single fluorescence intensity and fluorescence ratiometric optical fiber sensors. Also the NO intracellular release in the cytoplasm was done by the developed optical fiber fluorescence lifetime sensor. In Figure 2 is presented one example of the *in vivo* NO electrochemical sensing (a) and of the *in vivo* NO optical sensing (b) based in AuNPs previous described.

Some electrochemical amperometric sensors based in AuNPs were also developed for the *in vitro* sensing [35-37]. In the first one, hemoglobin is adsorbed with AuNPs [35], in another AuNPs were assembled with a sol-gel film [36] and, in the last electrochemical sensor, AuNPs were assembled in toluene/water interfaces [37].

4. Quantum Dots

The QD are crystalline semiconductor nanoparticles usually constituted by CdTe, CdSe, CdS, ZnS, CdTe/CdS, CdSe/CdS or CdSe/ZnS. They have a high fluorescence, narrow emission bands, large absorption spectra, large stokes shift, fluorescence tunable size, high surface to volume ratio and they are photostable. Also, varying the QD composition, it is possible to obtain NIR fluorescence emission using for example PbS, PbSe and InAs QD. Because of their composition some toxicity issues were posed. Due to their own optical properties they are essentially used for the optical fluorescence sensing and for imaging [38]. All of the developed NO sensors are based on fluorescence detection [39-50].

Only two of the developed sensors were used for *in vivo* NO sensing [39, 40]. Both sensors are based on a fluorescence quenching mechanism. In the first sensor, CdSe QD were encapsulated in modified hyperbranched polyether (mHP) nanospheres. Also, diazeniumdiolates (NO donors) were linked to the mHP-CdSe QD. The detection of the NO release in mouse fibroblast cells was done by the fluorescence quenching of NO donor-mHP-CdSe QD induced by the NO at an excitation wavelength of 480 nm with a charge coupled device (CCD) detector [39]. In the other sensor, CdTe QD were incorporated in carboxymethyl CS (CMCS). Beside the obtained CMCS-CdTe QDs, diazeniumdiolates (NO fluorescence probes) or O^2-substituted diazeniumdiolates (NO donors) were linked to the obtained CMCS-CdTe QD. The detection of the NO release by porcine iliac endothelial cells was performed by the fluorescence quenching of NO donor-CMCS-CdTe QD induced by the NO. A Stern-Volmer fluorescence quenching modelling was done using an excitation wavelength of 400 nm and emission of 560 nm with a linear range below 768 μM [40]. In Figure 3 is presented one example of the *in vivo* NO optical sensing (b) based in QD.

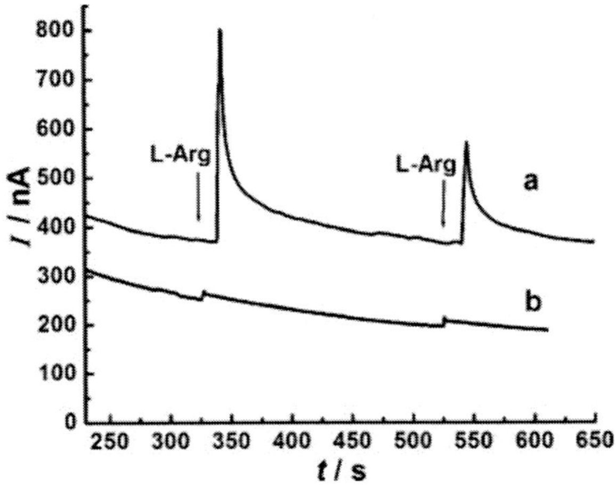

(a)

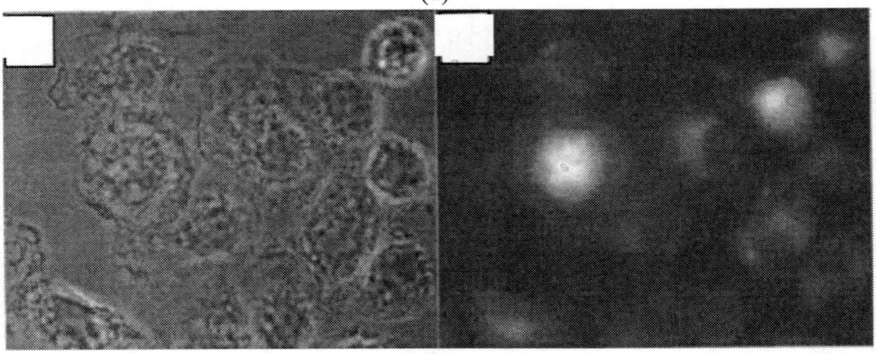

(b)

Figure 2. *In vivo* NO electrochemical sensing (a) by AuNPs-NA-Nafion electrode in a mice liver (reproduced from reference [30] with permission from Elsevier) and *in vivo* NO optical sensing (b) in macrophages cells by fluorophore labeled Cyt-c (reproduced from reference [32] with permission from the American Chemical Society).

Most of the developed NO optical sensors have potential to be used for *in vivo* NO sensing [41-50]. These are based in CdTe [44], CdSe [42, 43, 46, 49, 50] or in CdSe/ZnS QD [41, 45, 47, 48]. The response of these sensors is the fluorescence quenching induced by NO [42-47, 49]. Only for the CdSe QD [50] and CdSe/ZnS [41, 48] based sensors, a fluorescence enhancement is observed after quenching by a fluorescence resonance energy transfer (FRET) mechanism.

For the CdTe QD based sensor, the QD were dispersed in a poly(N-isopropylacrylamide) (PNIPAM) hybrid nanogel and diazeniumdiolates (NO donors) were posterior linked to the PNIPAM-CdTe QD [44]. For those sensors based on CdSe QD, and in order to stabilize the QD, they were immobilized in CS [42, 46]. In one of them, mercaptosuccinic acid (MSA) was linked to the CS-CdSe QD [42]. The other two were functionalized with triethanolamine (TEA) [49] and trivalent metalloporphyrins [50]. Also, a CdSe QD based optical fiber sensor uses cellulose as sensitive membrane [43]. In two of the developed NO optical sensors based on CdSe/ZnS QD they were dispersed in a polymethacrylate film after stabilization by hexadeylamine [45] and in an organic/inorganic hybrid organogels [47]. In the other two the CdSe/ZnS QD were functionalized with similar ferric dithiocarbamate compounds [41, 48].

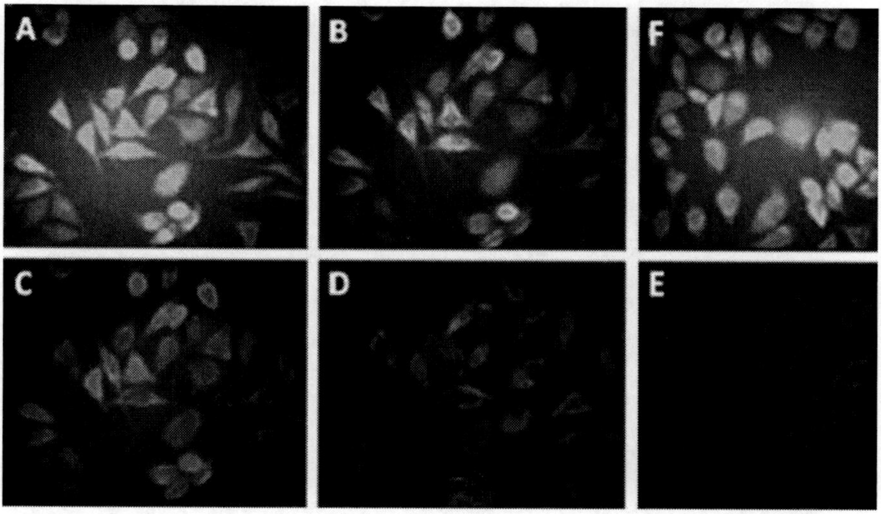

Figure 3. *In vivo* NO optical sensing in culture cells by a NO donor-CMCS-CdTe QD (reproduced from reference [40] with permission from Elsevier).

5. Other Nanomaterials

Beside the already indicated nanomaterials, others were also used for the *in vivo* sensing and/or imaging of NO. The most common nanomaterials families are metal oxides, carbon, silica and polymeric nanoparticles [11, 12]. The nanomaterials already used for the NO sensing are the metal oxide nanoparticles [51-54], silicon [55] and the different nanoallotropic forms of carbon [56, 57]. An *in vivo* NO electrochemical sensor [51] based in Al_2O_3

nanoparticles (Al_2O_3NPs) and an *in vivo* NO optical sensor [56] based in CD were developed.

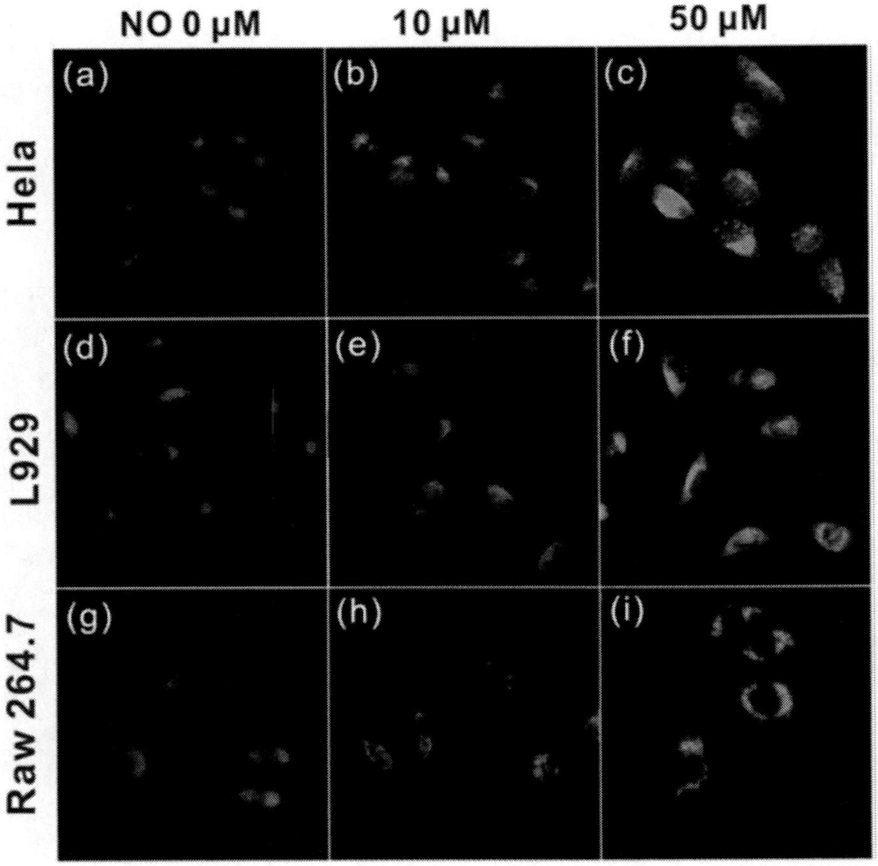

Figure 4. *In vivo* NO optical sensing by NOPD-CD in culture cells (reproduced from reference [56] with permission from the Royal Society of Chemistry).

In the developed *in vivo* NO electrochemical sensor an Al_2O_3NPs film was deposited at the surface of a GCE. With this electrode the detection of NO was achieved with a linear range of 0.04 – 210 μM, sensitivity of 0.037 μA/μM and a LOD of 7.2 nM. The NO detection in rat lung was done at an oxidation potential of + 0.69 V (vs. SCE) by amperometry [51]. In the developed *in vivo* NO optical sensor the CD were functionalized with naphthalimide with an o-phenylenediamine group (NOPD) by N-hydroxysulfosuccinimide sodium salt (Sulfo-NHS) / EDC cross-linking procedure. The fluorescence ratiometric NO sensing based in a FRET mechanism was done at an excitation wavelength of 370 nm and emission wavelengths of 455 and 535 nm. In the presence of NO there is a decrease of maximum fluorescence intensity at 455 nm and an increase at 535 nm. The ratio of maximum fluorescence intensity at 535 nm and 455 nm was evaluated in order to NO concentration. A detection limit of 3 nM with selectivity to NO_2^-; NO_3^-; ClO^-; $OH^{\cdot}$; H_2O_2; $^1O_2$ and ascorbic acid was evaluated. The detection of NO was done at different culture cells [56]. In Figure 4 is presented the *in vivo* NO optical sensing (b) based in CD previous described.

Mostly of the already developed *in vitro* NO sensors with potential for the *in vivo* NO sensing are amperometric electrochemical sensors [24,52-54,57]. Only one of these *in vitro* sensors is a fluorescence optical sensor [55]. The electrochemical sensors were based in Al_2O_3NPs modified with electropolymerized metallo 4′, 4″, 4‴, 4⁗tetra-amine phthalocyanine (poly-MTAPc) covered with nafion [52], in α-Fe_2O_3 nanoparticles (α-Fe_2O_3NPs) alone [53] or with CNT [24] embedded in a CS matrix, in SnO_2 nanoparticles (SnO_2NPs) film in which were immobilized Cyt-c and hemoglobin (Hb) [54] and in a GP nanosheet- hexadecyltrimethylammonium bromide (CTAB) / Hb-CS nanocomposite film [57]. The NO optical sensor was based in the fluorescence enhancement of silicon nanowires (SiNWs) functionalized with reduced fluoresceinamine [55].

Conclusion/Perspectives

The majority of the developed NO sensors based in nanomaterials are for *in vitro* determinations. The *in vivo* nanomaterials based NO sensors uses mainly carbon nanotubes and gold nanoparticles but, also, quantum dots, Al_2O_3 nanoparticles and carbon dots. Almost all the electrochemical sensors are amperometric sensors mainly based in the oxidation of NO. The fluorescent sensors are mainly based in the quenching induced by the NO.

Taking into consideration the characteristics of the nanomaterials under development and the enormous investment in their research, particularly in non toxic/biocompatible nanoparticles and in their functionalization, it is expected that nanomaterials could be increasingly used in the development of *in vivo* NO electrochemical or optical sensors with improved performances.

Acknowledgments

Financial support from Fundação para a Ciência e Tecnologia (FCT, Lisbon) project PEST and a PhD grant to Eliana Simões SFRH/BD/ 81074/2011 is acknowledged.

References

[1] Eckert, M.A., et al., Novel molecular and nanosensors for in vivo sensing. *Theranostics,* 2013. *3*(8): p. 583-94.

[2] Wilson, G.S. and M.A. Johnson, In-Vivo Electrochemistry: What Can We Learn about Living Systems? *Chemical Reviews,* 2008. *108*(7): p. 2462-2481.

[3] Wessels, J.T., et al., Advances in cellular, subcellular, and nanoscale imaging in vitro and *in vivo. Cytometry* A, 2010. *77*(7): p. 667-76.

[4] Rao, J., A. Dragulescu-Andrasi, and H. Yao, Fluorescence imaging in vivo: recent advances. *Current Opinion in Biotechnology*, 2007. *18*(1): p. 17-25.

[5] Paolo, S., Nitric Oxide in Human Health and Disease. *eLS,* 2013.

[6] Toledo, J.C., Jr. and O. Augusto, Connecting the chemical and biological properties of nitric oxide. *Chem Res Toxicol,* 2012. *25*(5): p. 975-89.

[7] Hetrick, E.M. and M.H. Schoenfisch, Analytical chemistry of nitric oxide. *Annu Rev Anal Chem* (Palo Alto Calif), 2009. *2*: p. 409-33.

[8] Yao, D., A.G. Vlessidis, and N.P. Evmiridis, Determination of Nitric Oxide in Biological Samples. *Microchimica Acta,* 2004. *147*(1-2).

[9] Xu, T., et al., Electrochemical Sensors for Nitric Oxide Detection in Biological Applications. *Electroanalysis,* 2014. *26*(3): p. 449-468.

[10] McQuade, L.E. and S.J. Lippard, Fluorescent probes to investigate nitric oxide and other reactive nitrogen species in biology (truncated form: fluorescent probes of reactive nitrogen species). *Curr Opin Chem Biol,* 2010. *14*(1): p. 43-9.

[11] Zamborini, F.P., L. Bao, and R. Dasari, Nanoparticles in measurement science. *Anal Chem,* 2012. *84*(2): p. 541-76.

[12] Valentini, F. and G. Palleschi, Nanomaterials and Analytical Chemistry. *Analytical Letters,* 2008. *41*(4): p. 479-520.

[13] Zhu, C., et al., Electrochemical Sensors and Biosensors Based on Nanomaterials and Nanostructures. *Analytical Chemistry,* 2015. *87*(1): p. 230-249.

[14] Ruedas-Rama, M.J., et al., Fluorescent nanoparticles for intracellular sensing: a review. *Anal Chim Acta,* 2012. *751*: p. 1-23.

[15] Moradi, O., et al., Carbon Nanotubes: A Review of Chemistry Principles and Reactions. *Fullerenes, Nanotubes and Carbon Nanostructures,* 2012. *20*(2): p. 138-151.

[16] Liu, Z., et al., Carbon Nanotubes in Biology and Medicine: In vitro and in vivo Detection, Imaging and Drug Delivery. *Nano Res,* 2009. *2*(2): p. 85-120.

[17] Santos, R.M., et al., Biomimetic sensor based on hemin/carbon nanotubes/chitosan modified microelectrode for nitric oxide measurement in the brain. *Biosens Bioelectron,* 2013. *44*: p. 152-9.

[18] Deng, X., F. Wang, and Z. Chen, A novel electrochemical sensor based on nano-structured film electrode for monitoring nitric oxide in living tissues. *Talanta,* 2010. *82*(4): p. 1218-24.

[19] Abdelwahab, A.A., et al., A selective nitric oxide nanocomposite biosensor based on direct electron transfer of microperoxidase: removal of interferences by co-immobilized enzymes. *Biosens Bioelectron,* 2010. *26*(3): p. 1080-6.

[20] Zheng, D., et al., Noncovalently functionalized water-soluble multiwall-nanotubes through azocarmine B and their application in nitric oxide sensor. *Electrochemistry Communications,* 2008. *10*(1): p. 90-94.

[21] Kim, J.-H., et al., The rational design of nitric oxide selectivity in single-walled carbon nanotube near-infrared fluorescence sensors for biological detection. *Nat Chem,* 2009. *1*(6): p. 473-481.

[22] Zhao, L., S. Zhu, and J. Zhou, A novel amperometric nitric oxide sensor based on imprinted microenvironments for controlling metal coordination. *Sensors and Actuators B: Chemical,* 2012. 171-172: p. 563-571.

[23] Porras Gutierrez, A., et al., Hybrid Materials from Carbon Nanotubes, Nickel Tetrasulfonated Phthalocyanine and Thin Polymer Layers for the Selective Electrochemical Activation of Nitric Oxide in Solution. *Electroanalysis,* 2009. *21*(21): p. 2303-2310.

[24] Gong, J., et al., Bifunctional sensor of nitric oxide and oxygen based on hematite nanotubes embedded in chitosan matrix. *Electrochemistry Communications,* 2008. *10*(9): p. 1222-1225.

[25] Zhang, J., et al., Single Molecule Detection of Nitric Oxide Enabled by d(AT)15 DNA Adsorbed to Near Infrared Fluorescent Single-Walled Carbon Nanotubes. *Journal of the American Chemical Society,* 2011. *133*(3): p. 567-581.

[26] Zhang, X., Gold Nanoparticles: Recent Advances in the Biomedical Applications. Cell Biochem Biophys, 2015.

[27] Yeh, Y.C., B. Creran, and V.M. Rotello, Gold nanoparticles: preparation, properties, and applications in bionanotechnology. *Nanoscale,* 2012. *4*(6): p. 1871-80.

[28] Zan, X., et al., Freestanding graphene paper decorated with 2D-assembly of Au@Pt nanoparticles as flexible biosensors to monitor live cell secretion of nitric oxide. *Biosens Bioelectron,* 2013. *49*: p. 71-8.

[29] Wang, F., et al., Nitric oxide measurement in biological and pharmaceutical samples by an electrochemical sensor. *Journal of Solid State Electrochemistry,* 2011. *15*(4): p. 829-836.

[30] Dang, X., et al., Gold nanoparticle film grown on quartz fiber and its application as a microsensor of nitric oxide. Sensors and Actuators B: Chemical, 2011. *160*(1): p. 260-265.

[31] Zhu, M., et al., Novel nitric oxide microsensor and its application to the study of smooth muscle cells. *Analytica Chimica Acta,* 2002. *455*(2): p. 199-206.

[32] Barker, S.L.R., et al., Ratiometric and Fluorescence-Lifetime-Based Biosensors Incorporating Cytochrome c ‘ and the Detection of Extra- and Intracellular Macrophage Nitric Oxide. *Analytical Chemistry,* 1999. *71*(9): p. 1767-1772.

[33] Barker, S.L.R. and R. Kopelman, Development and Cellular Applications of Fiber Optic Nitric Oxide Sensors Based on a Gold-Adsorbed Fluorophore. *Analytical Chemistry,* 1998. *70*(23): p. 4902-4906.

[34] Barker, S.L.R., et al., Fiber-Optic Nitric Oxide-Selective Biosensors and Nanosensors. *Analytical Chemistry,* 1998. *70*(5): p. 971-976.

[35] Xu, Y., C. Hu, and S. Hu, A reagentless nitric oxide biosensor based on the direct electrochemistry of hemoglobin adsorbed on the gold colloids modified carbon paste electrode. S*ensors and Actuators B: Chemical,* 2010. *148*(1): p. 253-258.

[36] Kannan, P. and S.A. John, Highly sensitive electrochemical determination of nitric oxide using fused spherical gold nanoparticles modified ITO electrode. *Electrochimica Acta,* 2010. *55*(10): p. 3497-3503.

[37] Li, Y.-J., et al., Large-scale self-assembly of hydrophilic gold nanoparticles at oil/water interface and their electro-oxidation for nitric oxide in solution. *Journal of Electroanalytical Chemistry,* 2008. *622*(1): p. 103-108.

[38] Kairdolf, B.A., et al., Semiconductor quantum dots for bioimaging and biodiagnostic applications. *Annu Rev Anal Chem (Palo Alto Calif),* 2013. *6*: p. 143-62.

[39] Liu, S., et al., Quantum dots-hyperbranched polyether hybrid nanospheres towards delivery and real-time detection of nitric oxide. *Mater Sci Eng C Mater Biol Appl,* 2014. *45*: p. 37-44.

[40] Tan, L., et al., Novel quantum dots-carboxymethyl chitosan nanocomposite nitric oxide donors capable of detecting release of nitric oxide in situ. *Acta Biomater,* 2012. *8*(10): p. 3744-53.

[41] Sun, J., et al., Fluorescence turn-on detection of gaseous nitric oxide using ferric dithiocarbamate complex functionalized quantum dots. *Anal Chem,* 2014. *86*(12): p. 5628-32.

[42] Simoes, E.F., J.M. Leitao, and J.C. Esteves da Silva, NO Fluorescence Quantification by Chitosan CdSe Quantum Dots Nanocomposites. *J Fluoresc,* 2013.

[43] Ding, L., et al., A sensitive optic fiber sensor based on CdSe QDs fluorophore for nitric oxide detection. *Sensors and Actuators B: Chemical,* 2013. *185*: p. 70-76.

[44] Tan, L., A. Wan, and H. Li, Quantum dots-poly(N-isopropylacrylamide) hybrid nanogel diazeniumdiolates as donors and real-time probes of nitric oxide. *Materials Chemistry and Physics,* 2013. *138*(2-3): p. 956-962.

[45] Fabregat, V., et al., Quantum dot–polymethacrylate composites for the analysis of NOx by fluorescence spectroscopy. *Inorganica Chimica Acta,* 2012. *381*: p. 212-217.

[46] Tan, L., et al., Biocompatible quantum dots–chitosan nanocomposites for fluorescence detection of nitric oxide. *Materials Chemistry and Physics,* 2012. *134*(2–3): p. 562-566.

[47] Wadhavane, P.D., et al., Organogel–quantum dots hybrid materials displaying fluorescence sensitivity and structural stability towards nitric oxide. *Soft Matter,* 2012. *8*(16): p. 4373.

[48] Wang, S., M.-Y. Han, and D. Huang, Nitric Oxide Switches on the Photoluminescence of Molecularly Engineered Quantum Dots. *Journal of the American Chemical Society,* 2009. *131*(33): p. 11692-11694.

[49] Yan, X.Q., et al., Fluorescence sensing of nitric oxide in aqueous solution by triethanolamine-modified CdSe quantum dots. *Luminescence,* 2009. *24*(4): p. 255-9.

[50] Ivanisevic, A., et al., Photoluminescent Properties of Cadmium Selenide in Contact with Solutions and Films of Metalloporphyrins: Nitric Oxide Sensing and Evidence for the Aversion of an Analyte to a Buried Semiconductor−Film Interface. *Journal of the American Chemical Society,* 2000. *122*(15): p. 3731-3738.

[51] He, Q., D. Zheng, and S. Hu, Development and application of a nano-alumina based nitric oxide sensor. *Microchimica Acta,* 2008. *164*(3-4): p. 459-464.

[52] Yap, C.M., G.Q. Xu, and S.G. Ang, Amperometric nitric oxide sensor based on nanoporous platinum phthalocyanine modified electrodes. *Anal Chem,* 2013. *85*(1): p. 107-13.

[53] Zhang, L., et al., Direct electrocatalytic oxidation of nitric oxide and reduction of hydrogen peroxide based on alpha-Fe2O3 nanoparticles-chitosan composite. *Talanta,* 2010. *82*(1): p. 196-201.

[54] Topoglidis, E., et al., Direct Electrochemistry and Nitric Oxide Interaction of Heme Proteins Adsorbed on Nanocrystalline Tin Oxide Electrodes. *Langmuir,* 2003. *19*(17): p. 6894-6900.

[55] Miao, R., et al., Modified silicon nanowires: a fluorescent nitric oxide biosensor with enhanced selectivity and stability. *Journal of Materials Chemistry,* 2012. *22*(8): p. 3348.

[56] Yu, C., et al., A fluorescent ratiometric nanosensor for detecting NO in aqueous media and imaging exogenous and endogenous NO in live cells. *Journal of Materials Chemistry* B, 2013. *1*(33): p. 4152.

[57] Wen, W., et al., A highly sensitive nitric oxide biosensor based on hemoglobin–chitosan/graphene–hexadecyltrimethylammonium bromide nanomatrix. *Sensors and Actuators B: Chemical,* 2012. *166-167*: p. 444-450.

In: Nitric Oxide
Editor: Dominic M. Evans
ISBN: 978-1-63482-961-8

Chapter 5

L-ARGININE METABOLISM IN THE PATHOGENIC TRYPANOSOMATIDS *TRYPANOSOMA CRUZI* AND *LEISHMANIA* SP.

R. F. Rodrigues[1], D. L. Fabrino[2] and L. L. Leon[1*]
[1]Laboratory of Biochemistry of Trypanosomatids,
Oswaldo Cruz Institute, FIOCRUZ, Brazil
[2]Department of Chemistry, Biotechnology
and Bioprocess Engineering, UFSJ/CAP, Brazil

ABSTRACT

Some trypanosomatids are associated with several diseases which are considered important public health problems, such as leishmaniasis, presenting in several clinical forms caused by different species of the genus *Leishmania*, and Chagas Disease, caused by *Trypanosoma cruzi*. During the infection, host cells produce toxic products, but trypanosomatids have a very well developed defense system that allows them to survive and proliferate within those cells. This system is based mainly in the L-arginine metabolism, since trypanosomatids are able to produce nitric oxide (NO) through the nitric oxide synthase (NOS) enzyme that uses L-arginine as a substrate. L-arginine is also a substrate for arginase (ARG) producing urea and polyamines, leading to the synthesis of trypanothione, a well-known molecule which is part of a redox system in

* Corresponding author: lleon@globo.com.

these parasites. The functions and interaction of these enzymes of the L-arginine pathway in the parasite could be related to their infectivity and virulence or their susceptibility to different drugs and free radicals. Furthermore, this pathway in the macrophages could determine the parasite death or its proliferation in a dependence of the intracellular availability of L-arginine.

The NO pathway has been described in *Trypanosoma cruzi* epimastigotes and is characterized by the enzyme NOS. Our research group demonstrated the NO production and characterized a cNOS from *L. amazonensis, L. braziliensis* and *L. infantum* (syn. *L. chagasi*) promastigotes and axenic amastigotes, NO production being more prominent in the latter.. The arginase gene was identified in *L. major* and *L. amazonensis. Leishmania* parasites have one arginase gene and data from studies with a mutant *L. mexicana* without arginase expression showed that these parasites were unable to synthesize polyamines besides presenting a decrease in their infectivity. However, until now no data concerning ARG have been described in *T. cruzi*. The polyamine biosynthesis in this case is obtained through another metabolic pathway, which is unique to *T. cruzi*. Recent data suggest a role of parasite L-arginine-/agamatine-derived polyamines in this process. It is important to point out the importance of polyamines and their relation with trypanothione biosynthesis, which is essential for parasite survival. As these trypanosamiases are highly pathogenic to humans and other mammals, and a neglected area of scientific research, despite being endemic to some South American countries, it is fundamental to understand their biochemistry in order to enable more specific interventions.

INTRODUCTION

Chagas disease and Leishmaniasis, caused by the protozoan parasites *Trypanosoma cruzi* and several *Leishmania* species respectively, are endemic in some countries in Latin America, constituting an important health problem. These are obligate intracellular parasites that infect cells of their vertebrate hosts, in which they exist as non-motile intracellular amastigotes. Within the host cells, these parasites, as a consequence of the respiratory burst, are exposed to oxygenated species, such as superoxide anion (O_2^-), hydrogen peroxide (H_2O_2) and the radical hydroxyl (OH^-). Also, the main nitrogenated radical produced by the host is NO, which is biosynthesized via NOS (Stamler et al., 1992). These molecules are individually toxic, and because of their high chemical reactivity, lead to the production of other radicals or molecules with potentially increased toxicity, such as peroxynitrite ($ONOO^-$). Protozoan parasites such as *T. cruzi* and *Leishmania* sp. have developed several strategies for evading the immune defense system of the host. Being eukaryotes, as are their host cells, they share several metabolic pathways, such as the L-arginine metabolism. For that reason, a comparative study of their similarities/ differences is important in order to define a target for the development of new molecules with potential anti-protozoan activity. This target must be a molecule or pathway which is essential for the parasite's survival. In fact, L-arginine is located in the intersection of divergent metabolic routes, producing metabolites that would have contrasting effect when evaluated in a parasite/host relation (Peluffo et al., 2004).

The most important mechanism for the killing of *Leishmania* and control of leishmaniasis is based on the role of L-arginine metabolism that can be modulated by Th1 and Th2 immune responses (Wanasen & Soong, 2008). Th1 cytokines such as interferon-γ (IFN-γ) and tumor necrosis factor-α (TNF-α), as well as the bacterial lipopolysaccharide (LPS), are responsible for the expression of an inducible form of NOS (iNOS) and the generation of large amounts of NO in neutrophils and macrophages (Stuehr & Marletta,1987). Under certain conditions, NO reacts with superoxide anions to form peroxynitrite, a potent oxidizing molecule which contributes to tissue injury during inflammatory processes (Bogdan et al., 2000). In contrast, Th2 response is characterized by high levels of the cytokines interleukin-4 (IL-4), IL-10, Il-13 which increase host cellular arginase activity resulting in alternative macrophage activation and enhanced production of polyamines for parasite growth and proliferation (Iniesta et al., 2001; 2002; Kropf et al., 2005).

The immune system can control *Leishmania* infection when a Th1 response is mounted and proinflammatory cytokines released, mainly by antigen-presenting cells (APC). This process leads to the induction of inducible nitric oxide synthase (iNOS) and NO production (Holzmuller et al., 2006). On the other hand, for parasites to survive within macrophages they depend on a Th2 response which modulates the arginase (ARG) expression. These enzymes, iNOS and ARG, have L-arginine as a common substrate for their activities modulated by both Th1 and Th2 cytokines, in a manner such that Th1 response increases conversion of arginine to NO, whereas the Th2 response promotes arginase induction (Wanasen & Soong, 2008). The inhibition of those enzymatic activities may be able to modulate macrophage infection by *Leishmania* sp. The *L. major* infection leads to ARG I induction in macrophages and that host enzyme induction supports *Leishmania* growth (Iniesta et al., 2005), while the increase of parasite ARG activity by IL-4 permits *Leishmania* survival within the host cell (Iniesta et al., 2001). Previous studies have demonstrated that the ARG activity in *Leishmania* sp. is able to modulate the NOS activity (Boucher et al., 1999; Wanasen & Soong, 2008). Moreover, studies using models of visceral leishmaniasis by *L. donovani* have shown that the parasite is able to modulate the activity of ARG in spleen cells of infected animals, increasing the enzyme expression and favoring the maintenance of infection. Strikingly, the active role played by *Leishmania* ARG in diverting arginine away from the iNOS pathway was demonstrated by the increased host NO response to ARG knockout *L. mexicana* infection in mice. The ARG knockout *L. mexicana* infection led to an enhanced Th1 response also associated to an INF-γ up-regulation (Reguera et al., 2009).

Considering the importance of the cellular signaling pathways involving the host cells and parasites of the *Leishmania* genus, there are few studies in the literature connecting the importance of the parasite ARG inhibition and macrophage NO production. Therefore, ARG-specific inhibitors may be considered a potential alternative among new substances with pharmacological interest for the treatment of leishmaniasis. *Leishmania* parasites reside within macrophages, and survival within this hostile environment hinges on their ability to modulate various macrophage-related immune functions, including alteration of reactive nitrogen species. The decrease of NO production in *Leishmania*-infected macrophages following inhibition of iNOS was reported by Balestrieri et al. in 2002. Although NO has great importance in cell signaling processes and is used by different organisms for this purpose, this molecule receives great scientific prominence in the literature as a relevant microbicide compound with a recognized role, mainly in innate immunity, including trypanosomatids. With regard to this parasite, NO is considered by many authors as a key molecule in the host immune response (Green et al., 1990a; 1990b; Evans et al., 1993; Assreuy et al., 1994; Murray & Nathan, 1999; Awasthi et al., 2004). However, different studies show that *Leishmania* has a remarkable ability to evade the host immune response and establish a chronic infection. Various pathogens, such as *Leishmania* sp., start their interaction with macrophages throughout the vertebrate host, cells of the innate immune system constituents which exhibit various defense mechanisms, including NO. However, such cells often have a double role in the response of the host's immunity against infection. On the other hand, NO is regarded as an important effector cell in the controlling and killing of microorganisms through oxidative and nonoxidative mechanisms. However, they can also serve as a place of lodging and multiplication for such microorganisms, protecting them from adverse substances of the extracellular environment or offering favorable microenvironmental conditions for their survival (Bogdan & Rollinghoff, 1999; Burchmore & Barrett, 2001; McConville et al., 2007). The amastigote form has unique ability to face the hostile environment of the endosomal compartment of macrophages and remain – albeit covertly, silently and in small numbers – for a long time (years) in the cells of their vertebrate hosts (Zambrano-Villa et al., 2002). Once in the endosomal compartment of vacuoles, one of the major toxic components of phagosomes beyond the oxygenate radicals is NO. However, the amastigote form of parasite in particular demonstrates the ability to survive in its host or capacity for suppressing production, resisting direct action of or producing its own NO (Genestra et al., 2003a; 2003b; Kima, 2007).

Furthermore, *Leishmania* acts on the metabolic pathway of L-arginine, which occupies a crucial amino acid position between two opposing pathways in the host cell and has great importance for metabolic parasite itself (Vincendeau et al., 2003; Kropf et al., 2005; Reguera et al., 2009). These parasites have a separate carrier, with high affinity for sodium (Na^+) and for L-arginine (Kandpal et al., 1995; Mishan-Shaked et al., 2006). Perhaps the parasite may be decreasing the amount of iNOS substrate for the host cell, since it is "stealing" the enzyme substrate from the host cell. This would lead to a decrease in the levels of NO produced by the induced NOS pathway. As a result, the parasite not only reduces harmful levels of NO produced by this route, but would benefit from this substrate to produce its own NO, acting also as a generator of the parasite's cyclic guanosine monophosphate (cGMP) (Peluffo et al., 2004).

The murine *L. major* infection is a well-studied model regarding the establishment of resistance or susceptibility to infection of the host, especially in macrophages, leading to the synthesis of NO which would be toxic to the parasite. In these cells, it is clear that the stimulation of Th1 cytokine components, mainly IFN-γ, leads to host cell resistance. However, the profile of Th2 cytokines will be capable of inhibiting NO production and stimulating the biochemical pathway of ARG, which uses the same substrate used in the production of NO. Thus, NO synthesis is down-regulated and, in parallel to producing L-arginine as final metabolites, polyamines are used for multiplication and growth of the parasite (reviewed in Vincendeau et al., 2003; Peluffo et al., 2004).

Arginases are also involved in the generation of polyamines, putrescine, spermidine, and spermine, which are important for the regulation of cell proliferation, cell differentiation and cell death (Iniesta et al., 2002). Macrophages may be able to express both ARG isoforms in response to LPS, although the literature is divided on this (Morris et al., 1998; Sonoki et al., 1997). These murine macrophages up-regulate the expression of ARG I, but not that of ARG II (Munder et al., 1999). On the other hand, inflammatory cytokines such as IFN-γ and TNF-α, alone or in combination with LPS, are able to induce iNOS. It has been suggested that the production of iNOS is usually accompanied by ARG II expression and that this ARG may contribute to buffering the overproduction of toxic NO (Cederbaum et al., 2004). Another way of inducing ARG I production is the stimulation of macrophages with Th2 cytokines.

L-arginine is an amino acid involved in several metabolic, physiological and pathologic processes. L-arginine has been well studied in mammals, with a focus on its participation in the urea cycle through the enzyme ARG and in the NO production mediated by the enzyme NOS. Concerning NOS activity, L-arginine goes to a two-step oxidation process, where initially there occurs a hydroxylation yielding N^{ω}-hydroxy-L-arginine (NOHA), which is then transformed in L-citruline and NO. Concerning ARG activity, L-arginine is hydrolized to L-ornithine and urea. This amino acid has been defined as essential to the growth of parasites of the genus *Leishmania,* which were not able to multiply in an L-arginine-deficient medium, suggesting a mechanism of transport and utilization of this molecule (Krassner & Flory, 1971; Camargo et al., 1978). This mechanism is carried out by an amino acid permease (LdAAP3) which was first described in *L. donovani* (Shaked-Mishan et al., 2006). These data show that these parasites as well as mammalian cells use L-arginine through a transport system (LdAAP3) and enzymes (NOS and ARG). There is an important inter-relation between those pathways, since L-ornithine is a substrate for ornithine decarboxylase (ODC) to synthesize polyamines, such as spermidine, which is directly related to the production of trypanothione. Similarly to glutathione in the mammal system, trypanothione is important both to the maintenance of the redox system in parasites and the host cell.

Although NO synthesis through NOS was preferentially studied in mammals, this route is not exclusive to these cells. In mammals there are three different isoforms of NOS which represent distinct gene products and constitute a NOS enzymatic family (reviewed by James, 1995; Alderton et al., 2001). Two isoforms are constitutive (cNOS), one described in neuronal cells (nNOS or NOS I) and the other in endothelial cells (eNOS or NOS III), as well as one inducible isoform (iNOS or NOS II). NO synthesis has also been observed in invertebrates (Johansson & Carlberg, 1995), always through the NOS enzyme. The first report of the NO pathway in an unicellular eukaryote was by Paveto and collaborators (1995), in *T. cruzi,* describing the presence of a constitutive NOS. The same cNOS was observed in *L. donovani*, but its biological significance, as well as the relation with the one present in the macrophages (inducible and hazardous to the parasite) remains to be defined (Basu et al., 1997). The isoforms described in *Leishmania* are similar to the constitutive one observed in mammals, concerning the same co-factors and the Ca^{2+} dependency. In 1997 the presence of a cNOS and NO production were demonstrated in *L. donovani*, but its function was not defined. Our group studied the importance of the NO production by the parasite and its survival within the macrophage, as well as the characterization of a cNOS and NO production by several *Leishmania* species. The effect of different inhibitors and a Ca^{2+} scavenger on the NO production by these parasites was also observed (Geigel, 2000; Geigel & Leon, 2003; Genestra et al., 2003a; 2003b).

The relation between the presence of a cNOS in *L. amazonensis* and its infective capacity was also observed, since the treatment of the parasite with L-nitro-arginine-methyl-ester (L-NAME), an inhibitor of cNOS, has an intense negative effect on this process (Genestra et al., 2003a; 2006). Concerning NO toxicity, several mechanisms of action have been described, such as the enzyme iron-dependent inactivation that could be a result of an association of NO to the prosthetic group iron and sulphur (Fe-S), producing a nitrosilation process on its active site (James, 1994; Oswald et al., 1994), or as peroxynitrite ($ONOO^-$) production, which is highly toxic to several cellular systems (Peluffo et al., 2004; Genestra, 2007). However, it is currently not clear what NO concentration could be considered toxic for one specific cell (Brunet, 2001). The concentration limit between the NO non-toxicity and the toxicity necessary to be a microbicidal molecule is very tenuous (Nagy et al., 2007). It is important to note that each cell/tissue system has its own unique resistance profile to NO (Kroncke et al., 1997). Furthermore, all organisms with the capacity to produce NO also have the potential to regulate its production, as well as to develop methods of detoxification and resistance to this radical. In mammals, the most important intracellular antioxidant is glutathione, which reacts with NO yielding S-nitrosoglutathione, resulting in the loss of NO oxidative potential. The depletion of this molecule in cell culture shows an increase in the susceptibility to the damage caused by NO, suggesting that pools of glutathione may act as NO scavengers (Kroncke et al., 1997; Genestra et al., 2008). In protozoa parasites, and especially in trypanosomatids such as *Leishmania* sp. and *T. cruzi*, there are no data about the identification of an enzyme associated with the neutralization of the reactive nitrogenated species.

Two isoforms of ARG (ARG I and II) have been described in mammal cells as being highly conserved, presenting structural and kinetic homologies in many biological systems but with different subcellular location, distribution, expression regulation and immunological reactivity (Jenkinson et al., 1996; Ash, 2004). ARG I is a cytosolic enzyme present mainly in the urea cycle, while ARG II is a mitochondrial protein potentially involved in proline, glutamate and polyamine synthesis.

Arginase expression in *Leishmania* was first suggested in research studying its enzymatic activity in parasite extracts (Camargo et al., 1978). The gene coding for ARG was identified in *L. major* and *L. amazonensis* (da Silva et al., 2002) and was cloned from *L. mexicana* promastigotes. *Leishmania* parasites contain an unique arginase gene (*ARG*), with biochemical properties and expression which have been studied in *L. amazonensis*, *L. mexicana* and *L. major* (da Silva et al., 2002; Reguera et al., 2009). The product of this gene is compartmentalized in glycosomes and is the main source of L-ornithine for polyamine synthesis in these parasites. Data from a *L. mexicana* strain knockout for *ARG* showed that these parasites were unable to synthesize polyamines, also observed in *L. major* (Reguera et al., 2009). Arginase from *L. mexicana* and *L. amazonensi*s were described within glycosomes, although this localization was not considered to be essential for the polyamine biosynthesis (da Silva et al., 2008). It is already known that ornithine decarboxylase (ODC) is considered the key enzyme for the *de novo* synthesis of polyamines (putrescine, spermidine, and spermine) in mammals and in some low eukaryotic organisms, such as *Leishmania* sp. However, in *T. cruzi* ARG activity has not been found and as a consequence the urea cycle is not functional (Opperdoes & Coombs, 2007)

Polyamines are policationic molecules found in animals, plants and microorganisms. The prominent polyamines in those biological systems are: the diamine putrescine, the triamine spermidine and the tetraamine spermine. As a result of their policationic nature, they can stabilize polymers rich in negative charges such as DNA, phospholipids and proteins. Spermidine, like the other polyamines (putrescine and spermine), is synthesized from methionine and ornithine, which are the amino acid precursors of polyamines in mammals (Wallace et al., 2003). The first step in the polyamine biosynthesis is the decarboxilation of the first amino acid precursor ornithine through the short half-life enzyme L-ornithine decarboxilase (ODC) producing putrescine, the decarboxilated ornithine (Pegg, 2006). In a parallel step, the second amino acid precursor methionine is activated to S-adenoylmethionine (SAM) which produces spermidine and spermine through decarboxilation by

S-adenosyl-methionine-decarboxilase (S-AdoMet DC). The final step consists in the conjugation of the decarboxilated ornithine and S-adenosine-decarboxilated methionine, by the enzyme propylaminotransferase I producing spermidine. The decarboxylases ODC and S-AdoMetDC are highly regulated in these mammal cells (Agostinelli et al., 2010).

The majority of eukaryotes have a polyamine transport system in the plasmatic membrane that allows the exogenous polyamine capture. In *L. donovani* ODC and S-AdoMetDC have long half-lives and this parasite has a very efficient polyamine transporter (POT1), which was also described in *L. major* (LmPOT1) (Hasne & Ullman, 2005). *T. cruzi* is unable to synthesize putrescine from ornithine, since ODC activity has not been detected (Cook et al., 2007). Although there are significant differences during its life cycle, most of the studies have been carried out in the epimastigote stage (Algranati, 2010).

Interestingly, *Trypanosoma cruzi* synthesizes polyamines by a different pathway and there is a controversy concerning how this parasite would do that. Biochemical studies have shown that *T. cruzi is* auxotrophic for polyamines, being unable to carry out the *de novo* biosynthesis of putrescine and therefore needing the addition of exogenous polyamines to the culture medium for its normal proliferation. *T. cruzi* epimastigotes lack ODC (Ariyanayagam & Fairlamb, 1997; Carrillo et al., 1999; 2003; Cook et al., 2007), and it is already known that the *T. cruzi* genome does not have genes encoding this enzyme (El-Sayed et al., 2005). Thus, *T. cruzi* requires diamine and/or polyamine transport mechanisms. Transport activities of putrescine and spermidine have been detected in intact *T. cruzi* epimastigotes (Le Quesne & Fairlamb, 1996; Gonzalez et al., 2001). This putrescine transport activity exhibits high affinity not only for putrescine but for cadaverine as well, but the transporter mechanism itself was not identified (Le Quesne & Fairlamb, 1996). The basic existing biochemical knowledge of polyamine transporters demands further studies not only in *T. cruzi*, but also in all eukaryotes (Gonzalez et al., 1992; Le Quesne & Fairlamb, 1996). Until recently, the only polyamine transporters that had been identified in eukaryotes were polyamine excretion proteins (Igarashi et al., 2001; Tachihara et al., 2005) or intracellular polyamine transporters (Tomitori et al., 2001) in *Saccharomyces cerevisiae*. Hasne and Ullman (2005) identified, as in *L.major,* two LmPOT1 orthologues within the *T. cruzi* genome, TcPOT1.1 and TcPOT1.2, each of which encodes a candidate polyamine permease and appears to be derived from distinct *T. cruzi* haplotypes. Furthermore, it was observed that both TcPOT1.1 and TcPOT1.2 are functional putrescine-cadaverine transporters, and that transporter function and location were directly regulated by the polyamine availability in the extracellular milieu.

Another important enzyme associated with polyamine synthesis, as an alternative pathway recognized in lower organisms, is the arginine decarboxylase (ADC) that also uses arginine as substrate, producing agmatine, which, through agmatinase (AGMAT) activity, will produce polyamines (Wang et al., 2014). The presence of ADC in trypanosomatids in general has been controversial for years, and in *T cruzi* specifically some very conflicting data have been published recently.

Molecular studies demonstrated that the wild-type *T. cruzi* genome does not contain ODC or ADC nucleic acid sequences, suggesting that the corresponding genes have been lost during evolution. It was observed that wild-type *T. cruzi* epimastigotes cultivated in a medium free of polyamines stopped growing after some passages, and the addition of arginine or ornithine was not efficient for re-initiation of parasite multiplication. On the other hand, normal growth was quickly restarted by adding exogenous putrescine or spermidine. These data indicate at least that there is no ADC enzymatic activity in this strain (Carrillo et al., 2003). It has been related that *T. cruzi* expresses very low but measurable ADC activity during logarithmic growth of epimastigotes, resulting in accumulation of polyamines and agmatine (Hernandez & Schwarcz de Tarlovsky, 1999; Piacenza et al., 2001). It was previously observed that the capacity of blood trypomastigote forms of *T. cruzi* to infect mouse peritoneal macrophages, as well the proliferation of intacellular amastigotes, was inhibited by treating the parasite with irreversible inhibitors of ADC, such as: DL-α-difluoromethylarginine (DFMA) or monofluoromethylagmatine. The inhibitory effect of DFMA, for instance, was significantly reduced by the presence of either agmatine or putrescine (Kierszenbaum et al., 1987; Yakubu et al., 1992). The parasites depend on putrescine uptake and S-adenosylmethionine decarboxylase (AdoMetDC) for their synthesis of the polyamines spermidine and spermine. The gene encoding AdoMetDC in *T. cruzi* was identified by PCR cloning, with degenerate primers corresponding to conserved amino acid sequences in AdoMetDC proteins of other trypanosomatids. The amplified DNA fragment was used as a probe to isolate the complete AdoMetDC gene from a *T. cruzi* genomic library. The protein showed a sequence identity of only 25% with human AdoMetDC, and, as expected, a higher sequence identity (68–72%) was found in comparison with trypanosomatid AdoMetDCs. Methylglyoxal-bis(guanylhydrazone) (MGBG), a potent inhibitor of human AdoMetDC, was a poor inhibitor of the *T. cruzi* enzyme (Personn et al., 1998). More recently, several researchers were unable to detect ADC activity in *T. cruzi* epimastigotes (Carrillo et al., 2003; Algranati, 2010). The impossibility of detecting the enzymes ODC and ADC in wild-type *T. cruzi* epimastigotes must be related to the absence of the corresponding genes in the parasite genome (Carrillo et al., 1999; 2004). Molecular studies suggest that *T. cruzi* is a natural deletion mutant of ODC and ADC genes. In an attempt to reverse the auxotrophy for polyamines, several experiments have been carried out transforming the parasites with plasmids containing those enzyme gene coding regions obtained from exogenous sources, as ODC from *Crithidia fasciculata* (Svensson et al.,1997) and from oat ADC cDNA (Bell & Malmberg, 1990). The transgenic parasites were able to express high levels of ODC and ADC. The one containing a foreign ODC gene became autotrophic for polyamines after acquiring the ability to synthesize putrescine and susceptible to DFMO. On the other hand, the one expressing ADC was not able to overcome polyamine auxotrophy (Carrillo et al., 1999; 2004; 2007).

The final step in this metabolic pathway is the synthesis of trypanothione, a thiol which, together with trypanothione reductase (TR), constitutes a very important redox system which neutralizes the hydrogen peroxide produced by macrophages during the parasite infection, allowing its survival within the host cell. Trypanothione [N^1-N^8-bis (glutathionyl) spermidine] is synthesized from the polyamine spermidine and gluthatione, through the enzyme glutathionyl spermidine synthetase (GES) yielding two glutathionyl spermidine molecules (N^1/N^8), which are coupled by the enzyme tripanothione synthetase (TS). Trypanothione was described for the first time by Fairlamb and Cerami (1985) in *T. brucei.* Since then, the molecule has been observed in many different biological systems, including *T. cruzi* (Krauth-Siegel et al., 1987; Jockers-Sherubl et al., 1989; Sullivan & Walsh, 1991 Carnieri et al., 1993) and *Leishmania* sp. (Keithly, 1989; Taylor et al., 1994; Cunnigham & Fairlamb, 1995; Dumas et al., 1997; Castro-Pinto et al., 2004; 2007; 2008) among others. Knockout parasites for TR are unable to neutralize the hydrogen peroxide, and as a consequence they will not survive inside the host cell.

CONCLUSION

In conclusion, based in the data presented here it is clear that further experiments must be done in order to get a better understanding of L-arginine metabolism, in particular in the *T. cruzi* protozoan parasite. It is important that a definition of the L-arginine metabolic pathway in both parasite and host cell be obtained, which will be useful for the development of potential chemotherapeutic agents against those pathologies. The search for a molecular target within the parasite remains a challenge for researchers.

REFERENCES

Agostinelli E, Marques MP, Calheiros R, Gil FP, Tempera G, Viceconte N, Battaglia V, Grancara S, Toninello A. 2010. Polyamines: fundamental characters in chemistry and biology. *Amino Acids*. 38(2), 393-403.

Algranati ID. (2010). Polyamine metabolism in *Trypanosoma cruzi*: studies on the expression and regulation of heterologous genes involved in polyamine biosynthesis. *Amino Acid*. 38, 645-651.

Alderton WK, Cooper CE, Knowles RG. (2001). Nitric oxide synthases: structure, function and inhibition. *Biochem. J.* 357, 593–615.

Ash DE. (2004). Structure and function of arginases. *J. Nutr*. 134: 2760S–2764S; discussion 2765S-2767S.

Ariyanayagam MR, Fairlamb AH. (1997). Diamine auxotrophy may be a universal feature of *Trypanosoma cruzi* epimastigotes. *Mol. Biochem. Parasitol*. 84,111-121.

Assreuy J, Cunha FQ, Epperlein M, Noronha-Dutra A, O'Donnell CA, Liew FY, Moncada S. (1994). Production of nitric oxide and superoxide by activated macrophages and killing of *Leishmania major*. *Eur. J. Immunol.* 24, 672-676.

Awasthi A, Mathur RK, Saha B. (2004). Immune response to *Leishmania* infection. *Indian J. Med. Res.* 119, 238-258.

Balestieri FM, Queiroz AR, Scavone C, Costa VM, Barral-Netto M, Abrahamsohn Ide A. (2002). *Leishmania (L.) amazonensis*-induced inhibition of nitric oxide synthesis in host macrophages. *Microbes Infect.* 4, 23-29.

Basu NK, Kole L, Ghosh A, Das PK. (1997). Isolation of a nitric oxide synthase from the protozoan parasite *Leishmania donovani*. *FEMS Microbiol. Lett.* 156, 43-47.

Bogdan C, Rollinghoff M. (1999). How do protozoan parasites survive inside macrophages? *Parasitol. Today*. 15, 22-28.

Bogdan C, Röllinghoff M, Diefenbach A. (2000). The role of nitric oxide in innate immunity. *Immunol. Rev*. 173, 17-26.

Brunet LR. (2001). Nitric oxide in parasitic infections. *Int. Immunopharmacol.* 1, 1457-1467.

Burchmore RJ, Barrett MP. (2001). Life in vacuoles-nutrient acquisition by *Leishmania* amastigotes. *Int. J. Parasitol.* 31, 1311-1320.

Camargo EP, Coelho JA, Moraes G, Figueiredo EN. (1978). *Trypanosoma sp.*, *Leishmania sp.* and *Leptomonas* sp.: enzymes of ornithine-arginine metabolism. *Exp. Parasitol.* 46, 141-144.

Carnieri EG, Moreno SN, Docampo R. (1993). Trypanothione-dependent peroxide metabolism in *Trypanosoma cruzi* different stages. *Mol. Biochem. Parasitol.* 61, 79-86.

Carrillo C, Cejas S, Gonzalez NS, Algranati ID. (1999). *Trypanosoma cruzi* epimastigotes lack ornithine decarboxylase but can express a foreign gene encoding this enzyme. *FEBS Lett.* 454, 192-196.

Carrillo C, Cejas S, Huber A, Gonzalez NS, Algranati ID. (2003). Lack of arginine decarboxylase in *Trypanosoma cruzi* epimastigotes. *J. Eukaryot. Microbiol.* 50, 312-316.

Carrillo C, Serra MP, Pereira CA, Huber A, Gonzalez NS, Algranati ID. (2004). Heterologous expression of a plant arginine decarboxylase gene in *Trypanosoma cruzi. Biochim. Biophys. Acta.* 1674, 223-230.

Castro-Pinto DB, Echevarria A, Genestra MS, Cysne-Finkelstein L, Leon LL. (2004). Trypanothione reductase activity is prominent in metacyclic promastigotes and axenic amastigotes of *Leishmania amazonensis*. Evaluation of its potential as a therapeutic target. *J. Enzyme Inhib. Med. Chem.* 19, 57-63.

Castro-Pinto DB, Lima EL, Cunha AS, Genestra M, De Leo RM, Monteiro F, Leon LL. (2007). *Leishmania amazonensis* trypanothione reductase: evaluation of the effect of glutathione analogs on parasite growth, infectivity and enzyme activity. *J. Enzyme Inhib. Med. Chem.* 22, 71-75.

Castro-Pinto DB, Genestra M, Menezes GB, Waghabi M, Goncalves A, De Nigris Del Cistia C, Sant'Anna CM, Leon LL, Mendonca-Lima L. (2008). Cloning and expression of trypanothione reductase from a New World *Leishmania* species. *Arch. Microbiol.* 189, 375-384.

Cederbaum SD, Yu H, Grody WW, Kern RM, Yoo P, Iyer RK. (2004). Arginases I and II: do their functions overlap? *Mol. Genet. Metab.* 81, S38-S44.

Cook T, Roos D, Morada M, Zhu G, Keithly JS, Feagin JE, Wu G, Yarlett N. (2007). Divergent polyamine metabolism in the apicomplexa. *Microbiology*. 153, 1123-1130.

Cunningham ML, Fairlamb AH. (1995). Trypanothione reductase from *Leishmania donovani*. Purification, characterisation and inhibition by trivalent antimonials. *Eur. J. Biochem.* 230, 460-468.

da Silva ER, Castilho TM, Pioker FC, Tomich de Paula Silva CH, Floeter-Winter LM. (2002). Genomic organisation and transcription characterisation of the gene encoding *Leishmania (Leishmania) amazonensis* arginase and its protein structure prediction. *Int. J. Parasitol.* 32, 727-737.

da Silva ER, da Silva MF, Fischer H, Mortara RA, Mayer MG, Framesqui K, Silber AM, Floeter-Winter LM. (2008). Biochemical and biophysical properties of a highly active recombinant arginase from *Leishmania (Leishmania) amazonensis* and subcellular localization of native enzyme. *Mol. Biochem. Parasitol.* 159, 104-111.

Dumas C, Ouellette M, Tovar J, Cunningham ML, Fairlamb AH, Tamar S, Olivier M, Papadopoulou B. (1997). Disruption of the trypanothione reductase gene of Leishmania decreases its ability to survive oxidative stress in macrophages. *EMBO J.* 16, 2590-2598.

El-Sayed NM, Myler PJ, Bartholomeu DC, Nilsson D, Aggarwal G, Tran AN, Ghedin E, Worthey EA, Delcher AL, Blandin G, Westenberger SJ, Caler E, Cerqueira GC, Branche C, Haas B, Anupama A, Arner E, Aslund L, Attipoe P, Bontempi E, Bringaud F, Burton P, Cadag E, Campbell DA, Carrington M, Crabtree J, Darban H, da Silveira JF, de Jong P, Edwards K, Englund PT, Fazelina G, Feldblyum T, Ferella M, Frasch AC, Gull K, Horn D, Hou L, Huang Y, Kindlund E, Klingbeil M, Kluge S, Koo H, Lacerda D, Levin MJ, Lorenzi H, Louie T, Machado CR, McCulloch R, McKenna A, Mizuno Y, Mottram JC, Nelson S, Ochaya S, Osoegawa K, Pai G, Parsons M, Pentony M, Pettersson U, Pop M, Ramirez JL, Rinta J, Robertson L, Salzberg SL, Sanchez DO, Seyler A, Sharma R, Shetty J, Simpson AJ, Sisk E, Tammi MT, Tarleton R, Teixeira S, Van Aken S, Vogt C, Ward PN, Wickstead B, Wortman J, White O, Fraser CM, Stuart KD, Andersson B. (2005). The genome sequence of *Trypanosoma cruzi*, etiologic agent of Chagas disease. *Science*. 309, 409-415.

Evans TG, Thai L, Granger DL, Hibbs JB, Jr. (1993). Effect of in vivo inhibition of nitric oxide production in murine leishmaniasis. *J. Immunol.* 151, 907-915.

Fairlamb AH, Cerami A. (1985). Identification of a novel, thiol-containing co-factor essential for glutathione reductase enzyme activity in trypanosomatids. *Mol. Biochem. Parasitol.* 14, 187-198.

Geigel LF. (2000). Estudo da via do óxido nítrico em *Leishmania amazonensis*. MSc Thesis. Rio de Janeiro, Fundação Instituto Oswaldo Cruz.

Geigel LF, Leon LL. (2003). Cyclic 3'-5' guanosine monophosphate-dependent activity in *Leishmania amazonensis*. *Mem. Inst. Oswaldo Cruz.* 98, 499-500.

Genestra M, Cysne-Finkelstein L, Guedes-Silva D, Leon LL. (2003a). Effect of L-arginine analogs and a calcium chelator on nitric oxide (NO) production by *Leishmania sp. J. Enzyme Inhib. Med. Chem.* 18, 445-452.

Genestra M, de Souza WJ, Cysne-Finkelstein L, Leon LL. (2003b). Comparative analysis of the nitric oxide production by *Leishmania sp.*. *Med. Microbiol. Immunol.* 192, 217-223.

Genestra M, Souza WJ, Guedes-Silva D, Machado GM, Cysne-Finkelstein L, Bezerra RJ, Monteiro F, Leon LL. (2006). Nitric oxide biosynthesis by *Leishmania amazonensis* promastigotes containing a high percentage of metacyclic forms. *Arch. Microbiol.* 185, 348-354.

Genestra M. (2007). Oxyl radicals, redox-sensitive signalling cascades and antioxidants. *Cell Signal*. 19, 1807-1819.

Genestra M, Soares-Bezerra RJ, Gomes-Silva L, Fabrino DL, Bellato-Santos T, Castro-Pinto DB, Canto-Cavalheiro MM, Leon LL. (2008). In vitro sodium nitroprusside-mediated toxicity towards *Leishmania amazonensis* promastigotes and axenic amastigotes. *Cell Biochem. Funct.* 26, 709-717.

Gonzalez NS, Ceriani C, Algranati ID. (1992). Differential regulation of putrescine uptake in *Trypanosoma cruzi* and other trypanosomatids. *Biochem. Biophys. Res. Commun*. 188, 120-125.

Gonzalez NS, Huber A, Algranati ID. (2001). Spermidine is essential for normal proliferation of trypanosomatid protozoa. *FEBS Lett.* 508, 323-326.

Green SJ, Crawford RM, Hockmeyer JT, Meltzer MS, Nacy CA. (1990a). *Leishmania major* amastigotes initiate the L-arginine-dependent killing mechanism in IFN-gamma-stimulated macrophages by induction of tumor necrosis factor-alpha. *J. Immunol.* 145, 4290-4297.

Green SJ, Meltzer MS, Hibbs JB, Jr., Nacy CA. (1990b). Activated macrophages destroy intracellular *Leishmania major* amastigotes by an L-arginine-dependent killing mechanism. *J. Immunol.* 144, 278-283.

Hasne MP, Ullman B. (2005). Identification and characterization of a polyamine permease from the protozoan parasite *Leishmania major*. *J. Biol. Chem*. 280, 15188-15194.

Hernández S, Schwarcz de Tarlovsky S. (1999). Arginine decarboxylase in *Trypanosoma cruzi*, characteristics and kinetic properties. *Cell Mol. Biol.* 45, 383-391.

Holzmuller P, Hide M, Sereno D, Lemesre JL. (2006). *Leishmania infantum* amastigotes resistant to nitric oxide cytotoxicity: Impact on in vitro parasite developmental cycle and metabolic enzyme activities. *Infect Genet Evol.* 6, 187-197

Igarashi K, Ito K, Kashiwagi K. (2001). Polyamine uptake systems in *Escherichia coli. Res. Microbiol.* 152, 271-278.

Iniesta V, Gómez-Nieto LC, Corraliza I. 2001. The inhibition of arginase by N(omega)-hydroxy-l-arginine controls the growth of *Leishmania* inside macrophages. *J. Exp. Med.* 193, 777-784.

Iniesta V, Gomez-Nieto LC, Molano I, Mohedano A, Carcelen J, Miron C, Alonso C, Corraliza I. (2002). Arginase I induction in macrophages, triggered by Th2-type cytokines, supports the growth of intracellular *Leishmania* parasites. *Parasite Immunol.* 24, 113-118.

Iniesta V, Carcelén J, Molano I, Peixoto PM, Redondo E, Parra P, Mangas M, Monroy I, Campo ML, Nieto CG, Corraliza I. (2005). Arginase I induction during *Leishmania major* infection mediates the development of disease. *Infect Immun.* 73, 6085-6090.

James SL. (1995). Role of nitric oxide in parasitic infections. *Microbiol. Rev.* 59, 533-547.

Jenkinson CP, Grody WW, Cederbaum SD. (1996). Comparative properties of arginases. *Comp. Biochem. Physiol. B. Biochem. Mol. Biol.* 114, 107-132.

Jockers-Scherubl MC, Schirmer RH, Krauth-Siegel RL. (1989). Trypanothione reductase from *Trypanosoma cruzi*. Catalytic properties of the enzyme and inhibition studies with trypanocidal compounds. *Eur. J. Biochem.* 180, 267-272.

Johansson KU, Carlberg M. (1995). NO-synthase: what can research on invertebrates add to what is already known? *Adv. Neuroimmunol.* 5, 431-442

Kandpal M, Fouce RB, Pal A, Guru PY, Tekwani BL. (1995). Kinetics and molecular characteristics of arginine transport by *Leishmania donovani* promastigotes. *Mol. Biochem. Parasitol.* 71, 193-201.

Keithly JS. (1989). Ornithine decarboxylase and trypanothione reductase genes in *Leishmania braziliensis guyanensis*. *J. Protozool.* 36, 498-501.

Kierszenbaum F, Wirth JJ, McCann PP, Sjoerdsma A. (1987). Arginine decarboxylase inhibitors reduce the capacity of *Trypanosoma cruzi* to infect and multiply in mammalian host cells. *Proc. Natl. Acad. Sci. USA* 84, 4278-4282.

Kima PE. (2007). The amastigote forms of *Leishmania* are experts at exploiting host cell processes to establish infection and persist. *Int. J. Parasitol.* 37, 1087-1096.

Krassner SM, Flory B. (1971). Essential amino acids in the culture of *Leishmania tarentolae. J. Parasitol.* 57, 917-920.

Krauth-Siegel RL, Enders B, Henderson GB, Fairlamb AH, Schirmer RH. (1987). Trypanothione reductase from *Trypanosoma cruzi.* Purification and characterization of the crystalline enzyme. *Eur. J. Biochem.* 164, 123-128.

Kroncke KD, Fehsel K, Kolb-Bachofen V. (1997). Nitric oxide: cytotoxicity versus cytoprotection-how, why, when, and where? *Nitric Oxide.* 1, 107-120.

Kropf P, Fuentes JM, Fahnrich E, Arpa L, Herath S, Weber V, Soler G, Celada A, Modolell M, Muller I. (2005). Arginase and polyamine synthesis are key factors in the regulation of experimental leishmaniasis in vivo. *FASEB J.* 19, 1000-1002

Le Quesne SA, Fairlamb AH. (1996). Regulation of a high-affinity diamine transportbsystem in *Trypanosoma cruzi* epimastigotes. *Biochem. J.* 316, 481-486.

McConville MJ, de Souza D, Saunders E, Likic VA, Naderer T. (2007). Living in a phagolysosome; metabolism of *Leishmania amastigotes. Trends Parasitol.* 23, 368-375.

Morris SM Jr, Kepka-Lenhart D, Chen LC. (1998). Differential regulation of arginases and inducible nitric oxide synthase in murine macrophage cells. *Am. J. Physiol.* 275, E740- E747.

Munder M, Eichmann K, Moran JM, Centeno F, Soler G, Modolell M. (1999). Th1/Th2-regulated expression of arginase isoforms in murine macrophages and dendritic cells. *J. Immunol.* 163, 3771-3777.

Murray HW, Nathan CF. (1999). Macrophage microbicidal mechanisms in vivo: reactive nitrogen versus oxygen intermediates in the killing of intracellular visceral *Leishmania donovani. J. Exp. Med.* 189, 741-746.

Nagy G, Clark JM, Buzas EI, Gorman CL, Cope AP. (2007). Nitric oxide, chronic inflammation and autoimmunity. *Immunol. Lett.* 111, 1-5.

Oswald IP, Wynn TA, Sher A, James SL. (1994). NO as an effector molecule of parasite killing: modulation of its synthesis by cytokines. *Comp. Biochem. Physiol. Pharmacol. Toxicol. Endocrinol.* 108, 11-18.

Paveto C, Pereira C, Espinosa J, Montagna AE, Farber M, Esteva M, Flawiá MM, Torres HN. (1995). The nitric oxide transduction pathway in *Trypanosoma cruzi. J. Biol. Chem.* 270, 16576-16579.

Pegg AE. 2006. Regulation ornithine decarboxylase. *J. Biol. Chem.* 281, 14529-14532.

Peluffo G, Piacenza L, Irigoin F, Alvarez MN, Radi R. (2004). L-arginine metabolism during interaction of *Trypanosoma cruzi* with host cells. *Trends Parasitol.* 20, 363-369.

Persson K, Aslund L, Grahn B, Hanke J, Heby O. (1998). *Trypanosoma cruzi* has not lost its *S*-adenosylmethionine decarboxylase: characterization of the gene and the encoded enzyme. *Biochem. J.* 333, 527–537.

Piacenza L, Pellufo G, Radi R. (2001). L-arginine-dependent suppression of apoptosis in *Trypanosoma cruzi:* Contribution of the nitric oxide and polyamine pathways. *PNAS* 98, 7301-7306.

Reguera RM, Balana-Fouce R, Showalter M, Hickerson S, Beverley SM. (2009). *Leishmania major* lacking arginase (ARG) are auxotrophic for polyamines but retain infectivity to susceptible BALB/c mice. *Mol. Biochem. Parasitol.* 165, 48-56.

Shaked-Mishan P, Suter-Grotemeyer M, Yoel-Almagor T, Holland N, Zilberstein D, Rentsch D. (2006). A novel high-affinity arginine transporter from the human parasitic protozoan *Leishmania donovani. Mol. Microbiol.* 60, 30-38.

Sonoki T, Nagasaki A, Gotoh T, Takiguchi M, Takeya M, Matsuzaki H, Mori M. (1997). Coinduction of nitric-oxide synthase and arginase I in cultured rat peritoneal macrophages and rat tissues in vivo by lipopolysaccharide. *J. Biol. Chem.* 272, 3689-3693.

Stamler JS, Singel DJ, Loscalzo J. (1992). Biochemistry of nitric oxide and its redox-activated forms. *Science.* 258, 1898-1902.

Stuehr DJ, Marletta MA. (1987). Induction of nitrite/nitrate synthesis in murine macrophages by BCG infection, lymphokines, or interferon-gamma. *J. Immunol.* 139, 518-525.

Sullivan FX, Walsh CT. (1991). Cloning, sequencing, overproduction and purification of trypanothione reductase from *Trypanosoma cruzi. Mol. Biochem. Parasitol.* 44, 145-147.

Tachihara K, Uemura T, Kashiwagi K, Igarashi K. (2005). Excretion of putrescine and spermidine by the protein encoded by YKL174c (TPO5) in *Saccharomyces cerevisiae. J. Biol. Chem.* 280, 12637-12642.

Taylor MC, Kelly JM, Chapman CJ, Fairlamb AH, Miles MA. (1994). The structure, organization, and expression of the *Leishmania donovani* gene encoding trypanothione reductase. *Mol. Biochem. Parasitol.* 64, 293-301.

Tomitori H, Kashiwagi K, Asakawa T, Kakinuma Y, Michael AJ, Igarashi K. (2001). Multiple polyamine transport systems on the vacuolar membrane in yeast. *Biochem. J.* 353, 681-688.

Vincendeau P, Gobert AP, Daulouede S, Moynet D, Mossalayi MD. (2003). Arginases in parasitic diseases. *Trends Parasitol.* 19, 9-12

Yakubu MA, Basso B, Kierszenbaum F. (1992). DL-alpha-difluoro-methylarginine inhibits intracellular *Trypanosoma cruzi* multiplication by affecting cell division but not trypomastigote-amastigote transformation. *J. Parasitol.* 78, 414-419.

Zambrano-Villa S, Rosales-Borjas D, Carrero JC, Ortiz-Ortiz L. (2002). How protozoan parasites evade the immune response. *Trends Parasitol.* 18, 272-278.

Wallace HM, Fraser AV, Hughes A. (2003). A perspective of polyamine metabolism. *Biochem. J.* 376, 1-14.

Wanasen N, Soong L. (2008). L-arginine metabolism and its impact on host immunity against *Leishmania* infection. *Immunol. Res.* 41, 15-25.

Wang X, Ying W, Dunlap KA, Satterfield MC, Burghardt RC, Wu G, Bazer FW. (2014). Arginine decarboxilase and agmatinase: an alternative pathway for the de novo biosynthesis of polyamines for the development of mammalian conceptuses. *Biol. Reprod.* 90, 84: 1-15.

INDEX

A

B

C

D

E

F

G

H

I

N

O

P

Q

R

S

T

U

V

W

Y

Z